Life in the Universe
A Beginner's Guide

ONEWORLD BEGINNER'S GUIDES combine an original, inventive, and engaging approach with expert analysis on subjects ranging from art and history to religion and politics, and everything in-between. Innovative and affordable, books in the series are perfect for anyone curious about the way the world works and the big ideas of our time.

Life in the Universe
A Beginner's Guide

Lewis Dartnell

ONEWORLD

LIFE IN THE UNIVERSE

A Oneworld Book
Published by Oneworld Publications 2007
Reprinted 2009, 2012, 2015

Copyright © Lewis Dartnell 2007

All rights reserved
Copyright under Berne Convention
A CIP record for this title is available
from the British Library

ISBN: 978–1–85168–505–9

Typeset by Jayvee, Trivandrum, India
Cover design by Two Associates
Printed and bound in Great Britain by Clays Ltd, St Ives plc

Oneworld Publications
10 Bloomsbury Street
London WC1B 3SR
England

Stay up to date with the latest books,
special offers, and exclusive content from
Oneworld with our monthly newsletter

Sign up on our website
www.oneworld-publications.com

For Vicky
I love you

Contents

Preface

You hold in your hands *Life in the Universe: A Beginner's Guide*; thank you for picking it from the shelf. This book tells the story of our place in the cosmos and the prospects for finding life beyond the Earth, a field of scientific research that has come to be called 'astrobiology'. It has been written as both a popular science book, for anyone curious about the existence of life 'out there', and also an entry-level primer for undergraduate or postgraduate students starting a course in astrobiology. A glossary and reading list have been included at the back of this book to help those starting afresh in this field.

Over the coming chapters we'll consider some of the most fundamental questions in science. What actually is 'life'? How did it develop on our own world? In what way might the Earth be special, or the entire Universe be fortuitously set-up to allow life? Where else in the solar system or galaxy as a whole might the conditions for life be met? We'll take a look at some of the hardiest lifeforms known on Earth, how cells might be transferred between planets, and what dangers lurk out in the depths of space. We'll travel through four billion years of our planet's history and take a guided tour of the mostly likely abodes in our own solar system, before voyaging out to the pinprick stars that speckle our skies.

I encountered three main problems writing this book on astrobiology. The first was both a joy and a frustration. As astrobiology is such a fast-paced discipline, some information will have become out-dated even before the ink has dried on the page. For example, since completing the draft the first terrestrial planet orbiting another 'main sequence' star has been discovered, a world only five times bigger than Earth. The planet is not thought able to bear life as it lies far beyond the habitable zone of its cool red dwarf star, but it does constitute the most Earth-like discovery so far. An equally ground-breaking finding has been made within our own solar system. Enceladus, a small moon of Saturn, has been discovered to be surprisingly active; spewing out a plume of water from an area of blue 'tiger stripes' near its south pole. Astrobiologists have always over-looked this unimposing moon, but the realization now is that with pockets of liquid water there is the possibility of life even here. In support of this is the steadily amassing evidence that certain cold-tolerant cells on Earth remain active well below $-20\ ^{\circ}$C, the temperature previously thought to represent the lowest limit for life.

Second, this new discipline spans an enormous diversity of different research areas, each with their own extensive body of background knowledge and terminology. I have done my best to present all these facets equally, whilst being careful not to swamp the text with excessive jargon or peripheral detail.

Thirdly, astrobiology is still at an early stage of development, striving to secure its foundations in scientific bedrock. Much of the data collected so far is ambiguous and we have a very incomplete picture of many of the most crucial areas, resulting in fierce controversy and debate over correct interpretations. I have tried to fairly present all sides of an argument, but been careful not to lose the reader in an intricate and bewildering web of conflicting evidence, claims and counter-arguments. Despite discussing many of these ideas with the leaders in the field and rigorously checking every fact, misrepresentations and factual errors will still invariably have crept into the text. For these I apologise.

It goes without saying that this book would have been impossible without the selfless efforts of many people, offering me their time to comment on the evolving drafts and help check facts and theories, as well as countless fascinating conversations in conference coffee rooms, corridors and staircases. A few deserving special mention include Alan Aylward, Emily Baldwin, Tom Bell, Andrew Coates, Ian Crawford, Chris Lintott, Nicola McLoughlin, John Parnell, Andrew Pomiankowski, Dave Waltham, John Ward, and Julian Wimpenny. Great thanks also to my expert illustrator Piran Sucindran, copy editor Ann Grand and the hard-working team at Oneworld Publications; Marsha Filion, Kate Smith, and Mike Harpley.

But perhaps the most important acknowledgements are for those who don't realise how valuable their influence has been. I speak here of my friends and family that have hauled me through the rough patches of this project, and especially my grandfather who bought a young schoolboy his first popular science book, and opened my mind to the marvels of the Universe. I owe you all an incalculable debt of gratitude.

Lewis Dartnell
London, 2007

List of Illustrations

Introduction

As I write this, I occasionally look up and gaze on the world resting on my desk. I don't mean a satellite photo stuck to the wall or a globe balanced on a bookshelf. This world is a complete living, breathing, swarming, thriving system, the condensed essence of Earth, encapsulated in a hollow glass globe no more than six inches across; an entire ecosystem on a Lilliputian scale, sealed from the outside world, self-contained, self-regulated and with a potentially infinite lifespan. The idea behind my 'EcoSphere' is elegantly simple. The four main components of the Earth: lithosphere, hydrosphere, atmosphere and biosphere (rock, water, air and life), are reproduced in miniature within the glass orb. Land is provided by a handful of pebbles on the bottom of the ball, a few inches of water recreate the ocean and a bubble of atmosphere is trapped above. On a warm day water evaporates from the surface of the pool and condenses, like a cloud, on the underside of the glass. Drops form and rain back into the ocean, completing the water cycle. Animal, plant and bacterial life forms are contained within the microcosm, represented by a small group of shrimp, green algae and a horde of invisible microbes clinging to the surface of the pebbles or free-floating through the ocean. Nothing can enter or leave; all the materials necessary for life must cycle endlessly within the system. The miniature world's inhabitants depend on each other for survival but the algae are the keystone. The bubble of life depends on them, as they use sunlight to produce the nutritious sugars and oxygen the shrimp and some bacteria need to survive. The whole system is driven by light; energy released by nuclear fusion reactions ninety-three million miles away in the Sun's broiling core.

As well as being wonderfully distracting, this fragile little orb perfectly demonstrates many concepts that are crucial to life on Earth and therefore what we can deduce about the possibility of organisms on other worlds. This set of questions about our own origins, what processes occurred during the development of life, what conditions or raw materials are necessary and where else these prerequisites might be satisfied is shaping itself into a new field of science. Astrobiology, the study of life among the stars, is one of the hottest areas of multi-disciplinary research, fusing knowledge from biology, chemistry, astrophysics and geology. This multi-disciplinary study of the 'origin, evolution, distribution and future of life in the Universe' is sometimes called exobiology, xenobiology or bioastronomy. This book, a beginner's guide to astrobiology, will take us on a tour through the most exciting lines of thought in this new science, the latest findings and what still remains to be understood. By way of introducing some of the most important ideas, for the moment I'll turn back towards my toy bubble world, my EcoSphere.

Within the EcoSphere, some things are very obviously alive. The shrimp actively swim around, feed off the algae, grow, and occasionally reproduce to create another individual in their own likeness. If one were to be captured and sucked into an analysis tube, it could be seen to be reducing the levels of oxygen in the water, while simultaneously releasing carbon dioxide. If the shrimp is denied oxygen or nutrients for long enough, it will cease to display this activity and will be said to be dead. On the other hand, the pebbles resting on the bottom of the EcoSphere are clearly non-living. They do not move, respond to their environment, grow or divide and are completely inert as far as the levels of chemicals in the water are concerned. Drawing, from our observations, the conclusion that shrimps are alive while pebbles are dead, is simple enough in this example. But what about observations made by one of our robotic probes sent to an alien world to search for life? How would we know what signs to look for? What chemical processes might betray its activity? How can we be so sure that life, completely independent of our own, which has followed its own evolutionary course for billions of years, would be anything like us? Would we recognise alien life if we landed right on top of it? What makes us think it would be carbon-based and living in water and not built from completely different molecules and employing an exotic biochemistry? For that matter, what actually is 'life'?

The EcoSphere also demonstrates the distinction between two fundamentally different walks of life. Some organisms are self-sufficient and can support growth by extracting raw materials and energy from their environment; others consume other organisms. The algae are *photosynthetic*; using the energy from sunlight to make complex biomolecules to feed and re-create themselves (photosynthetic literally means 'building with light'). The algae are self-sufficient but the shrimp and bacteria are entirely reliant on the nutrition they provide. One of the greatest unknowns regarding the origin of life on Earth is whether the first cells fed on pre-formed organic molecules or were completely self-sufficient. Other than photosynthesis, what other sources of energy might alien life capitalise on? Could photosynthesis be a common trick for life, soaking up the light of suns throughout the galaxy? Oxygen released by photosynthesis has built up to a high level in the Earth's atmosphere and such a feature may also indicate the existence of similar life on distant planets, a signature we could detect light years away. The evolution of oxygen-releasing photosynthesis also produced one of the most profound changes in the history of the Earth.

The EcoSphere is completely sealed; any materials needed by life must circulate within the system. Both the shrimp and the algae produce carbon dioxide as a waste product, which is a raw material for the algae's photosynthesis. The carbon cycles invisibly within the sphere; released by the shrimp, stored as gas in the atmosphere, dissolving into the water, taken up by the algal photo-synthesis and locked into the complex molecules used to build its cells, eaten by the shrimp, released again and so on round the circle of life. This mirrors the passage of carbon through the hydrosphere, atmosphere and biosphere of the Earth. But the little EcoSphere misses one other crucial aspect of the real full-sized ecosystem – carbon can also become locked in rocks and carried deep into the Earth's interior by plate tectonics, before being released in volcanic eruptions. The rocky lithosphere constitutes a crucial fourth stage in the carbon cycle. The smooth running of the carbon cycle and the involvement of plate tectonics is thought by some scientists to have been crucial for the long-term stability of the Earth's biosphere. Every organism on Earth is not only inextricable linked to many others through food webs but also into the very fabric of the planet, with the essential elements circulating though rock, water, atmosphere and biosphere.

The EcoSphere, with its continually-circulating elements sealed within a transparent barrier, is a closed system, identical to the Earth isolated in space. This is true in terms of matter; energy, in the form of sunlight, must come from outside. As long as the EcoSphere receives a steady supply of light it could potentially run indefinitely, just as a watermill could turn perpetually in a river. But of course my little ecosystem won't actually live forever. Such a small dynamic network is particularly sensitive to perturbations: too much of a disturbance away from its healthy equilibrium could cause it to 'crash'. If the globe is left in the sun all day, the water may warm too much, heated by the greenhouse effect of the all-enclosing glass wall, and the shrimp die. Too great a period of darkness and the photosynthetic algae die. Exactly the same dangers threaten our planet: the steadily brightening Sun will eventually boil off the oceans and sterilise the Earth's surface. There is evidence in the fossil record of long stretches of time when almost all photosynthetic activity ceased, the Sun's light blocked out by a global layer of thick ice. In the microcosm trapped in my glass sphere, the death of the algae and thus the disappearance of the only source of both oxygen and food would almost immediately lead to the shrimps falling extinct. With the disappearance of the algae and the shrimp, the countless hordes of bacteria suspended in the water or clinging to the pebbles would be living on borrowed time. The complex organic molecules of the dead higher organisms would fuel bacterial growth and division for a limited period before this last component of the collapsing ecosystem would also starve to death. Chemically, the entire system will run out of available energy, and decay to a stable, but dead, equilibrium. Eventually the complex structures of the bacterial cells would break apart and all that would be left of the once vibrantly dynamic ecosystem would be an inert watery soup of basic organic compounds. In one sense, 'life' is nothing more than a self-sustaining entity, complex enough to use the available energy to maintain its own complexity and eventually reproduce to beget more.

My EcoSphere is a working model of an Earth-like planet; one with an oxygen atmosphere and great oceans of water. But the young Earth was dramatically different. How did the primordial, hell-like, Earth develop into the cool, wet place we now call home? What range of conditions can different forms of terrestrial life survive? We shall look into hardy cells that thrive in boiling acid or within pockets of saturated salt water trapped in solid ice. All the life within the EcoSphere ultimately depends on light-catching algae but in later chapters we shall explore entire communities which survive in the dark depths of the oceans and bacteria miles underground, which eat rock and are completely independent of the Sun's energy. Could similar cells be surviving in niches in the solar system? We shall tour other prime possible astrobiological locations in our own backyard; visit aquifers of liquid water deep beneath Mars' rusty surface, float among the clouds high in Venus's atmosphere and drift through the frigid global ocean trapped under the face of Europa. What about worlds even further afield, orbiting the pinprick stars scattered across the heavens?

Astronomers are discovering new planets at a prodigious rate – which of these are the most likely to be inhabited? What is a Red Dwarf (when it's not a sci-fi comedy series)? Could photosynthetic organisms be soaking up the rays of a dim red alien star? Might cells survive even in outer space? Could life itself be transferred between planets in our solar system or even between the stars, like a cosmic infection spreading through the galaxy?

Astrobiology

The field of astrobiology encompasses many things. It is the study of diverse 'extreme' organisms on Earth, research into the origins of life on our own planet and discussion on what processes and environmental conditions might be necessary. It is the search for potential niches for life, both within our solar system and throughout the galaxy. It is the design of lab experiments into prebiotic chemistry or into alternatives to the idiosyncratic systems developed by terrestrial life. It is the construction of rigorous life-detecting experiments and instruments, flown on space probes throughout the solar system. My particular field of research is into the levels of radiation which any life surviving on Mars would need to endure and this tiny piece of the jigsaw puzzle would fall into place somewhere on page 129.

Astrobiology is necessarily multi-disciplinary and to some extent it is difficult to define the field of knowledge it covers. Some of it is speculative, or based on incomplete data, which is to be expected of a fledgling discipline. Some critics have described it as a science that has yet to demonstrate its subject matter even exists, in that not one single example of life beyond Earth is known. This is undeniably true, but not in the sense of doggedly searching for a mystical beast like the Loch Ness Monster. As with any science, hypotheses are postulated, experiments constructed, theories refined according to the data and improved experiments designed.

Astrobiology was not born to explain a particular discovery, as microbiology developed to explain sightings through the first microscopes but is the offspring of more mature scientific disciplines as they overlap. Astrobiology can point to no specific inception but has gradually gathered momentum and acceptance over the last fifty years, as discoveries within biology and space exploration show that extra-terrestrial life really is possibile.

I shall begin the story with the most fundamental question of all: what actually is life?

1

The workings of life

What is life?

What is life? This, seemingly straightforward, question has plagued biologists, philosophers and theologians for centuries. It is mindlessly easy to discriminate between what is alive and what is dead. Jaguars, oaks and mushrooms are obviously living. Some life is recognised on a different time- or size-scale. Watched over a long enough period, lichen, the coloured crusts on old stone walls, can be seen to grow and develop; a chemical test of the air around them would tell you they were indeed photosynthesising. Magnified fifty times through a microscope, a drop of pond water teems with minuscule organisms. On the other hand, rocks, fires and clouds are clearly non-living. However, simply providing a list of things which are alive is not the same as being able concisely to define what are the properties of 'life'. Schoolchildren are often taught that life is defined by a checklist of seven characteristics: that living things eat, excrete, move, grow, reproduce, respond to changes in their environment and maintain a constant internal state. Some non-living entities satisfy a few of these attributes: fires grow and spread, self-sustaining from a flow of energy, consuming fuel and excreting waste products by the same oxidation reactions that run a cell and the ordered pattern of atoms in a crystal is able to reproduce itself. Equally, some living things do not tick all of the boxes: mules are sterile and unable to reproduce; although each of its component cells is alive, the whole animal fails the seven-part test. We need a more sophisticated definition.

Life as information transmission

One attempt to characterise life is known as the *Darwinian definition*. First, this states that life must contain a description of itself; an operating manual or set of instructions on how it can be rebuilt. Crystals are thus excluded, as they do not contain a true description of themselves but grow only because their structure organises free units onto an existing pattern. Second, it states that the individual must be able to carry out the instructions on its own and so self-replicate. This rules out viruses, as they reproduce by hijacking the molecular machinery of the host cell they have infected. All life on Earth has its operating manual: its text, a set of genes within the DNA molecule. A huge array of proteins translates and performs the instructions. This type of classification is therefore also known as a genetic definition. (Genetic here has a much broader sense than just 'section of DNA'; something much closer to its Greek root meaning 'mode of formation'.) Third, it states the system must be capable of evolution by natural selection. This implies that the method of duplicating genetic information should be inaccurate, so that errors or mutations are introduced, creating random variation within a population of replicators, ensuring that when faced with environmental stress only some survive to reproduce. This is the mechanism of Darwinian evolution; over time the replicators adapt to become better suited to their surroundings. This process honed the abilities of the first replicating molecules to produce cells, animals and eventually, almost four billion years later, the self-aware species that we are. It is often said that the human body is nothing more than an elaborate organic robot, solely designed to aid the replication of our DNA.

The Darwinian definition prescribes that life need only possess a system of information storage and transference – transmitting the genetic instructions to the next generation. Life is defined by what it *does*, not what it is *made of*. This classification is much less restrictive than others and includes 'non-biological' life. The development of artificial or A-life, is a burgeoning field: many different systems have been built with, for example, replicating computer code replacing organic polymers and hard discs the primordial soup. The processes of mutation, competition, death and evolution are the same, only the supporting medium is different.

Life as energy disequilibrium

A second definition for life specifies that in addition to information transmission, the system must extract energy to maintain itself. The problem is that a self-replicating system demands an extraordinary level of complexity. Complex organisations are very improbable – there are vastly more disordered ways to arrange a cloud of atoms than ordered into a functioning cell. Everything in the Universe naturally deteriorates from a state of high order to one more disarrayed. In technical terms, systems fall down the entropic gradient, from an ordered state of low entropy, to an equilibrium level with greater entropy. Life constantly fights this trend towards degeneration, keeping itself far from equilibrium. It does this by pumping in energy – it takes work to maintain an ordered state. Energy can be extracted from something as it degenerates; for example the heat given off by an ordered wooden log as a rush of oxidation reactions reduces it to ash and hot gas. Effectively, life allows one system to slide down the slope of organisation to push another uphill. A fungus sprouting from a tree stump survives by extracting the same energy as a fire but in a carefully controlled manner. Life requires a constant energy flow and can only survive where there is an external gradient. A little later in this chapter we will see how different forms of terrestrial life extract energy from their surroundings.

Modern terrestrial life performs these functions elegantly. It holds a complete description of itself as well as an elaborate network of chemical reactions which release energy, harnessing it to build useful molecules and maintain its own complexity. An army of proteins oversees this metabolic network and provides the machinery to carry out the instructions contained in the DNA and to copy it for the next generation. A third attribute of terrestrial life is that it is contained within an enclosed space. All life on Earth is cellular; bound by a membrane which physically separates the inside from the outside, preventing the different components from simply drifting apart, allowing control over the internal situation, the import and hoarding of valuable nutrients, the exclusion of waste products and the creation of chemical gradients to allow energy generation. Once, information storage and metabolic reactions were separate; in Chapter 4 we shall explore theories as to how these two crucial functions came to be integrated into the first cell. But first, what exactly is a cell?

Cells

Traditionally, there were thought to be only two fundamentally different forms of life on our planet. Animal cells like our own, as well as those of plants and fungi, store their DNA within a nucleus and are called *eukaryotes*. Bacteria represent another, more ancient, form of life without a nucleus, called *prokaryotic* (literally meaning 'before the nucleus'). Their DNA, twisted into a closed loop, floats free within the watery cytoplasm of the cell. Eukaryotic DNA is stabilised into chromosomes, within the nucleus, and this innovation seems to have allowed a much greater information capacity – the eukaryotic genome is up to 10,000 times larger than a bacterium's.

This is not the only difference between the two cell types. In addition to the nucleus, the inside of an eukaryote is extensively subdivided into many different compartments or 'organelles', such as *mitochondria*, the 'powerhouses' of the cell, which perform many of the reactions that extract energy from food compounds to

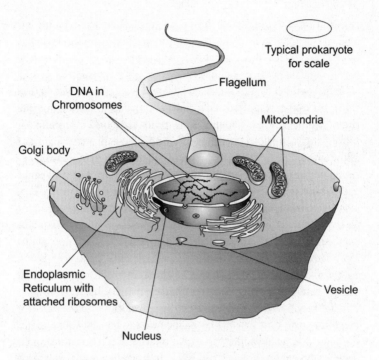

Typical prokaryote for scale

Flagellum

DNA in Chromosomes

Mitochondria

Golgi body

Endoplasmic Reticulum with attached ribosomes

Vesicle

Nucleus

Figure 1 Schematic cut-away of a typical eukaryotic animal cell. The defining feature of such a complex cell is the chromosomal DNA within the nucleus, but other crucial innovations include organelles like mitochondria and the Golgi body.

produce the energy-storing molecule *adenosine triphosphate* (ATP) and *chloroplasts*, present in algal and plant cells, which perform the light-harvesting reactions of photosynthesis. The process of protein synthesis is also much more sophisticated in eukaryotes. The ribosomes that make proteins are similar to the prokaryotic version but are concentrated on a special membrane which billows out from the nucleus. From there, new proteins are passed to a stack of organelles, where they are further processed, step-by-step, into a finished product. Figure 1 shows a rough schematic of a typical animal cell. Plant cells, in addition to the specialised features shown, have a thick cell wall for protection and support, a large central sac for storage and a horde of chloroplasts for photosynthesis.

Eukaryotic cells also have a far more advanced scaffolding and transport network, with thick strands of proteins strengthening the outer membrane or forming long poles that other proteins or whole organelles can be dragged along. These protein filaments allow a much more refined process of cell division than in prokaryotes. They also give the eukaryotic cell superb control over its outer membrane, which it can use to crawl over the surface or predate on smaller cells to engulf and digest them. This final ability is thought to have been an extremely important development, as it explains how eukaryotic cells acquired their mitochondria and chloroplasts. Both these organelles are believed to have once been free-living bacteria. The evidence for this is abundant: mitochondria and chloroplasts show many features characteristic of bacteria, such as circular DNA strands and sensitivity to antibiotics, and they are not created by the host cell but reproduce themselves. At some point, they became swallowed by an early eukaryotic cell but not broken down. Over evolutionary time, the bacterium and its host cell became so inter-dependent that they are now quite inseparable. The nucleated cell is absolutely dependent on the energy provided by its mitochondria (and chloroplasts for photosynthetic eukaryotes); in return the organelles receive essential nutrients and protection. Such a close partnership is known as *symbiosis*; when one organism actually lives inside the other it is called *endosymbiosis*.

To contain this internal organisation, eukaryotic cells are generally much larger than prokaryotic. The total number of prokaryotic cells living on and in the human body outnumbers, by at least ten times, the amount of eukaryotic cells: you're more bacterial than you are human. There is an enormous difference in scale between the smallest bacteria and the largest eukaryote. There are nanobacteria just 0.4 micrometres across, barely large enough to contain all the molecular machinery thought crucial for life; the largest cell known is a fist-sized eukaryote, a giant amoeba found in ocean abysses that builds itself a protective cage of sediment. By comparison, the largest bacterium is just over half a millimetre in size – only just visible to the naked eye but a Goliath of the microbial world.

This traditional understanding (that there were two fundamental forms of life – cells with a nucleus and cells without) was the story until recently. The evolutionary history of mammals can be reconstructed from distinctive features, such as traits of their skeletons, traced back through the fossil record but similar techniques are impossible for the minute features of single cells. However, in the early 1980s, a new technique based on the sequence of letters in a particular gene was used to draw the family tree of all life on Earth. This gene codes for a small subunit of the ribosome, the tiny structure that is involved in translating the DNA code into proteins. This is such a fundamental process that parts of the ribosome gene are essentially the same for all organisms, so sequence changes between the different gene versions can be used to calculate an organism's closest relatives. The results were astounding. As expected, the eukaryotes, the cells with a nucleus, cluster on one great branch of the tree of life. The shock is that the prokaryotes are not a single group but are split into two great domains, the bacteria and the *archaea*. There is as much genetic difference between human cells and the germ that gives us a stomach upset as there is between it and an archaea living in a hot spring. The root of this great tree, the hub central to all three bushy domains, is also fascinating. It tells us which living cells are closest to the universal ancestor, the organism from which all life on Earth is descended. But for that part of the story you'll have to wait until Chapter 4.

The true tree of life is certainly much messier than the one commonly depicted. Different cells often swap genes with each other, something they did much more back in evolutionary time, enormously complicating the issue of working out how they are related. Moreover, the evolution of the eukaryotes involved the wholesale importing of entire bacteria to form mitochondria and chloroplasts. It is not certain whether the original host cell was a bacteria or archaea but extensive gene studies show that eukaryotes contain far more bacteria-like material than archaeal. The eukaryotes bridge the two other domains, transforming the tree into more of a 'ring of life'.

The molecular workings of life

A cell can be summed up as a 'membrane-bound chemical reactor and information storage system'. But how is matter animated to produce life? What molecules are responsible for these amazing abilities? The three main components of a cell are the outer membrane, the genetic system and the metabolism.

The cell membrane is formed from a layer of molecules with 'head' ends that dissolve easily in water and fatty 'tail' regions that do not. Arrays of these molecules spontaneously arrange themselves to keep their fatty tails out of water, forming into a double layer with the tails aligned end-to-end in the middle. Metabolism is the general name given to the staggeringly complex network of reactions which interconvert the chemicals within a cell. Many of the small organic molecules (to a chemist, 'organic' simply means molecules containing carbon) are thought to have existed naturally from non-biotic processes on the early Earth, becoming incorporated into metabolism as it developed and expanded. Over time, life began to invent new, larger, molecules to serve its functions. For example, long chains of sugar subunits, such as starch, are a compact way of stockpiling energy and carbon. Other sugar polymers give rigid support; chitin is used to build the hard outer skeletons of arthropods (such as insects and crabs) and bacteria are protected by a wall of amino sugar polymers. Cotton shirts, wooden bookshelves and the pages of this book are all mostly cellulose fibres, a plant polymer built of units of glucose. However, for sheer diversity of structures and roles within the cell, the ultimate polymers are proteins. They are used for everything, from structural support, transporting valuable chemicals, sending signals and as enzymes accelerating the metabolic reactions. The subunits of proteins are amino acids, a class of small molecules with acid and alkaline ends and a side group that gives each amino acid slightly different properties. With only a few exceptions, all proteins on Earth are constructed using combinations of twenty amino acids. Some side groups dissolve well in water, while others are insoluble. For a protein to be soluble, its long chain must curl up into a complicated three-dimensional shape, held together by bonds between the side groups, which protects the insoluble amino acids on the inside. The finished shape is extremely precise; proteins are able to recognise and bind to other proteins, DNA, RNA or metabolic reactants with exquisite specificity. One class of proteins, enzymes, are custom-made tools that catalyse reactions (enable the chemical processes to occur much more rapidly than they would otherwise). Each enzyme can only make one small change to a compound; perhaps sticking on a phosphate group, removing an hydroxyl unit or breaking a bond to snap a carbon chain in two. Complete metabolic pathways contain huge numbers of different enzymes which gradually transform one molecule into another.

The stability of enzymes and other soluble proteins is a key issue in biology. As enzymes are warmed, the amino acid strands vibrate; at high temperatures, they vibrate so much that the bonds between side groups are broken and the protein loses its vital shape. If water gets inside as well, the entire protein structure is destroyed and falls out of solution. This is called *denaturing* – and can be seen when you fry an egg, as the albumin protein solidifies and turns white. Changing the electrical charge around a protein, as in extreme salinity or acidity, can also denature it. (As when you use a lemon juice marinade to 'cook' raw fish.) In the next chapter we'll look at organisms that thrive in the extreme environments of boiling hot springs, saturated salt solutions or highly acidic water and see what coping strategies they've evolved.

Another important aspect of the molecules used by cells is their 'handedness'. Any non-symmetrical shape possesses a quality called *chirality*. To see it for yourself, put down this book for a second and hold out your two hands, palm side up. Although your hands are essentially identical shapes, it is impossible to orient one on to the other. They are mirror images; no amount of rotating or flipping your right hand can make it look like your left hand. The same is true of most molecules in a cell; they have a particular handedness or chirality. The two possible mirror image versions of a molecule are called *enantiomers* and are as different as your left and right hands. Intriguingly, all terrestrial life uses just one enantiomer. All biologically-produced amino acids are left-handed, as shown in Figure 2, whereas all sugars are right-handed. However, the molecules produced in labs which simulate the prebiotic

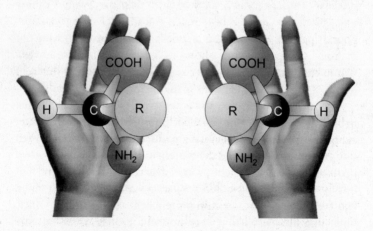

Figure 2 The chirality of amino acids. Only the left–handed version is used by terrestrial life. 'R' here denotes the side group, of which 20 different kinds are used by natural proteins.

chemistry of the early Earth are always an even mixture of both enantiomers. Why life only bothers with one enantiomer is easy to understand – the enzymes running metabolism are so specific that they can only work on their target molecule if it is the correct chirality. It would be enormously wasteful to have left- and right-handed versions of every enzyme, so life is stuck with one. The real mystery is what caused one enantiomer to be selected over the other in the first place.

For astrobiologists, the predominance of one enantiomer is a very good biosignature of alien life. For the Mars probes, instruments have been designed to test the chirality of any organic molecules that might be found.

The two most fundamental requirements for life are a coded operating manual and the ability to extract energy from the environment. How are these two basic functions performed? The mechanism for storing, translating and copying the genetic information is the more universal, so I'll start there.

Information storage

The cell's operating manual is encoded in the molecule *deoxyribonucleic acid* (DNA). DNA is a polymer, a long string of nucleotide subunits bonded together. Each nucleotide is made up of three parts, a sugar, a base and a phosphate group. The sugar, the deoxyribose, is a pentagon; four carbon atoms and one oxygen linked into a ring. The base can be one of four kinds: adenine, guanine, thymine and cytosine (abbreviated to A, G, T and C), each with a skeleton of nitrogen and carbon atoms. These are the letters of the genetic code. The backbone of the polymer is formed from sugar molecules linked together by phosphate groups, with the bases sticking out to the side. Two strands aligned together bind if their sequence of bases matches; A always pairs with T and G with C. This forms the famous double helical structure of DNA: two strands coiled around each other linked, like the rungs of a twisted ladder, by bases. This elegant structure is the key both to the information storage capabilities and the reproductive abilities of these molecules. The sequence of genetic letters specifies how to build proteins and the ladder construction of the DNA means that it can act as a template for its own replication. All three domains of cells share the same genetic code, so it must have developed before they diverged. This is the best evidence we have that all life on Earth has the same root.

To copy the genetic information during cell division, a co-ordinated team of enzymes is needed. Some force the two strands apart, unwinding the helix as they go, others attach spare nucleotides to the exposed bases and others join them together to form two complete new strands.

A similar process produces a temporary copy of the information to build a specific protein; the DNA helix is similarly unzipped but only along the appropriate gene and it is not DNA units that are base-paired and sealed together but *ribonucleic acid* (RNA) units. RNA is virtually identical to DNA, except that it has an extra oxygen atom attached to the pentagonal sugar ring and uracil (U) replaces thymine. Thus, the gene stored as DNA is transcribed into a messenger RNA molecule (mRNA), which makes its way to a ribosome. Ribosomes are themselves partly made from RNA and perform the crucial cellular job of translating the code in the mRNA into a protein. mRNA code is made up of triplets of bases – each set of three specifying a single amino acid. Spare amino acids are attached to a transfer RNA (tRNA) molecule bearing the appropriate triplet. The ribosome reads the mRNA message and attaches amino acids one after the other, recognising them by their tRNA tag, to create the particular protein. The protein then folds up into its own precise three-dimensional shape and is held together by bonds between the different amino acids.

To summarise, DNA stores the information on how to make the parts of a cell and proteins copy it every time the cell divides.

By itself, DNA cannot produce proteins. RNA is the middle-man, conveying the information from the DNA to the machinery that builds proteins. The importance of this interplay between DNA, RNA and proteins will become apparent in Chapter 4, when we'll consider the emergence of life. Proteins perform another, equally important role in the cell, generating useful energy.

Energy extraction

A fungus extracts energy from a tree stump by oxidising the complex molecules that make up wood. Oxidation, as a chemical process, not only involves adding oxygen to a molecule (or removing hydrogen) but also stripping an electron from it. Reduction is the reverse reaction, whereby an electron is donated to a molecule. These reactions always occur in pairs: an electron is taken from one molecule (the *reductant* or fuel) and passed to another, the *oxidant*. This reduction–oxidation processes is often abbreviated to *redox*. All cells in the human body; indeed almost all eukaryotes extract energy by breaking down a carbohydrate molecule (such as the sugar glucose) and oxidising it to produce water and carbon dioxide. Humans need to breathe oxygen because their cells use this molecule as the final electron acceptor: electrons are stripped from the carbohydrate, passed along a chain of intermediate molecules to perform work for the cell and ultimately given to oxygen to produce water. This process, *respiration*, occurs in the mitochondria within human cells.

Respiration

A chemical's degree of oxidation is measured by its *redox potential*: strong oxidants, such as oxygen gas, have large positive values; strong reductants, such as glucose, have very negative ones. How much energy a particular reaction will yield is indicated by the difference between the redox potentials of the oxidant and the reductant. An electrical battery works in exactly the same way; the negative terminal is attached to a reductant and the positive to an oxidant. When a wire joins the two, a redox reaction proceeds and the current of electrons can be used to do work. As seen in Figure 3, the complete oxidation of glucose yields a lot of energy; cells must slowly break down the molecule step-by-step to release a little energy at a time. Glucose, a ring of six carbon atoms, is converted into molecules of two carbons. Along the way, some of the energy-storage molecule, ATP, is made and a few electrons are captured by electron carrier molecules. The two-carbon compound is now fed into a great circular chain of reactions that exists in all eukaryotes. It is called the Krebs cycle, and at the beginning the two-carbon compound is added to one with four carbons. This molecule is steadily oxidised, releasing an electron here, producing ATP there, snipping off a carbon here, until the original four-carbon compound is regenerated and the cycle begins again.

The Krebs cycle can turn indefinitely, churning out electrons as it goes. Some of these energetic liberated electrons are used to drive the building of complex reduced molecules, pushing them up the hill of organisation that I mentioned earlier in the chapter but most are used to produce ATP, in a very ingenious mechanism. The electrons are taken to the inner wall of the mitochondria by the carriers and deposited at the start of a long electron transport chain. Once on the chain, the electrons are passed from one component to the next, in a series of redox reactions, with many of the steps resulting in a proton (a hydrogen atom stripped of its electrons) being pushed outside the membrane, creating a proton gradient. Each component is slightly lower in energy state, until eventually the electron is thrown away to the final electron acceptor (oxygen, in human cells).

Before too long, there is a high concentration of protons

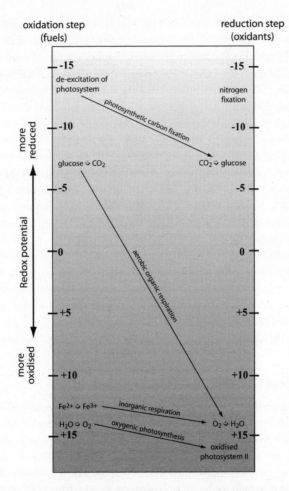

Figure 3 The ladder of redox couples. Life can only extract energy from reactions that proceed 'downhill', such as reacting glucose or Fe^{2+} with oxygen. Energy must be pumped in to the system to drive negative reactions, such as fixing carbon into glucose (e.g. light energy in photosynthesis) or nitrogen fixation. Splitting water to release oxygen requires enormous oxidising power, as provided by photosystem II in cyanobacteria and chloroplasts.

outside the membrane. The cell has effectively converted the energy of the glucose into a chemical gradient, in the same way that electricity companies pump water uphill into reservoirs as an energy store. Throughout the membrane there are specialised enzymes called ATP synthase. They allow the protons to flow back down the gradient, using their energy to drive the manufacture of ATP, like the rush of dammed water spinning a turbine. By using the electron transport chain, the Krebs cycle provides as much as three-quarters of a eukaryote's total energy demand. The overall efficiency of *aerobic* (carried out using oxygen) respiration of glucose has been calculated at around forty per cent – comparable to a coal-fired power station. Complete oxidation needs oxygen to accept the electrons at the end of the transport chain; without it, the electron carriers cannot be refreshed and the Krebs cycle grinds to a halt. Some eukaryotes and many bacteria respire *anaerobically* (without oxygen) in the process of fermentation, which releases about a tenth of the stored energy. In fermentation, carbohydrates are partially oxidised, to the point of entry to the Krebs cycle, and then converted to waste products (such as alcohol in the case of brewer's yeast) and expelled from the cell.

Inorganic energy

If the speciality of the eukaryotes is their genetic control and co-ordination, assembling trillion-cell constructs like the human body, the prokaryotes' forte is their metabolic diversity. Eukaryotes can, in general, only feed on things they can directly pass to the Krebs cycle (although fungi are a little less finicky, feasting on methanol, for example). But bacteria and archaea are true chemical wizards, capable of living off almost every imaginable food source.

In principle, any redox reaction yields electrons that can be used to drive cellular metabolism. For example, ferrous iron (Fe^{2+}), an ion with a charge of $+2$ (ions are atoms or molecules that carry an electrical charge, due to an imbalance of electrons) can be oxidised to give a ferric ion with a charge of +3 and a free electron: $Fe^{2+} \rightarrow Fe^{3+} + e^-$

Ferrous iron is green; when it loses an electron and becomes Fe^{3+} it turns red–brown, the distinctive colour of rust, Martian rock and human blood. If something with a higher redox potential than ferrous iron is present, which can act as the final electron acceptor, the whole reaction can be run like a biological battery, with the flow of electrons diverted down the transport chain to generate ATP and power the cell's activities. The fundamental position of the electron carriers and the transport chain in the biochemistry of all cells implies that they probably formed the basis of the very first metabolism. A few of the possible redox pairs are shown in Figure 3. Some prokaryotes use oxygen as the final electron acceptor and are called *aerobes*. Many do not, and can even be poisoned by the destructive effects of this powerful oxidising agent, so they are termed *anaerobes*.

Some prokaryotes employ heavy metal ions, such as arsenic, as the final electron acceptor; ions other cells find poisonous. Some bacteria even reduce uranium ions to run their metabolism. It is possible that very early life may have indirectly used nuclear power to drive their biochemistry: the abundance of radioactive isotopes of uranium, thorium and potassium would have been much higher on the young Earth than today. These unstable atoms release radiation when they decay, which can give an energy boost to surrounding molecules, so they react together. Some scientists believe that a substantial amount of reduced organic molecules could have been produced in this way and metabolised by life in the rocks. Such cells would have effectively been feeding off nuclear reactions.

Thus, many prokaryotes are able to derive their metabolic energy from inorganic reactions – they don't need to break down complex organic molecules like glucose. Some don't even need access to organic molecules as a carbon source for their cellular building blocks (amino acids, sugars, bases and so on). They are completely self-sufficient and can build everything they need from scratch, turning basic compounds like carbon dioxide into all the biological molecules they require.

Organisms can be classified by these two factors: their requirements for energy and for carbon source, as shown below in Table 1.

Carbon source	
organic	*hetero-*
CO_2	*auto-*
Energy Source	
organic electron donor	*organo-*
inorganic electron donor	*litho-* / *chemo-*
sunlight	*photo-*

For example, a human cell, which needs everything handed to it on a plate, is an *organoheterotroph*, a bacterium that survives on inorganic ion reactions but needs an external supply of sugars is a *lithoheterotroph*, while an archaea that uses inorganic substances not only for its energy demands but also as its carbon source is a *chemoautotroph*. Converting inorganic carbon dioxide into organic molecules, pushing them up the energy gradient shown in Figure 3, is known as *carbon fixation*. Some of the chemoautotrophs that do this live deep underground, extracting the ions and carbon dioxide they need from the minerals that surround them; quite literally eating rocks for breakfast. We'll look at some of these astounding organisms in the next chapter, when we discuss extremophiles. Many of the potential niches for life in the solar system are far from the Sun and devoid of pools of ready-made organic compounds, so chemoautotrophs are the odds-on favourites for extra-terrestrial life.

The table above shows a third energy source. *Photoautotrophs* are organisms that use the energy of sunlight to run their metabolism and fix carbon dioxide into carbohydrates. This process, *photosynthesis*, is enormously important in Earth's biosphere.

Photosynthesis

Like respiration, the key to photosynthesis is the liberation of electrons to perform work for the cell. Photosynthesis has two essential parts; the first captures light energy and the second uses that energy to drive carbon fixation. Central to the first part is a class of magnesium-containing compounds, the *chlorophylls*. These molecules are extremely efficient at absorbing visible light, using its energy to excite electrons within their structure. These electrons are picked up by carriers and shuttled to an electron transport chain.

Modern plants have two photosystems for trapping light. The first is a cyclic pathway, where the released electrons are run down a transport chain before returning to the chlorophyll. This is an ancient mechanism; a similar system provides energy for many photosynthetic bacteria. In the second photosystem, light energy also excites electrons, sending them down a transport chain or storing them in a carrier to drive later reduction reactions, but it also causes the splitting of a molecule of water. The low-energy electron released by this splitting is used to replace the one lost by the photosystem. The other by-product, oxygen, escapes as waste. This process is therefore called *oxygenic*, as it produces oxygen gas. In both photosystems, the electron transport chain creates a proton gradient that is used to generate ATP, just as in respiration.

The second part of photosynthesis – using the captured energy to build organic molecules – does not require light. The rubisco enzyme attaches a carbon atom (from carbon dioxide) to a five-carbon sugar, in a crucial carbon fixation step. This six-carbon sugar then breaks in half, to give two three-carbon molecules that enter the start of a great reaction loop, effectively the Krebs cycle in reverse. ATP and the carriers generated from the photosystems pump energy and electrons into the cycle, driving the reactions. Some of the products are siphoned off as activated six-carbon sugars to create all the complex carbohydrates that a plant needs. At the end, the five-carbon sugar is re-created and the cycle turns again.

Overall, photosynthesis generates ATP and fixes carbon dioxide into carbohydrates, gaining the electrons it needs by splitting water to release oxygen. Thus, both the fundamental demands of the cell, biochemical energy and reduced carbon compounds, are simultaneously satisfied. The evolution of photosynthesis, and the effect it had on the atmosphere, was utterly crucial in the history of life on Earth, a development I shall delve into more deeply in Chapter 4. Today, Earth's photosynthetic organisms fix about ten thousand billion tonnes of carbon every year, supporting the entire surface biosphere. Rubisco is one of the most abundant enzymes on the entire planet and leaves a distinctive signature in the carbon it fixes, a biosignature which can be used, in ancient rocks, to date the emergence of oxygenic photosynthesis, as we'll see in Chapter 4.

The Earth's crust has a natural negative redox potential; the atmosphere and oceans are slightly more oxidised, by the action of sunlight and radiation. This creates an electrochemical gradient (or *potential difference*) between the crust and the water and air above it – a planet-sized battery. Many lithotrophs extract energy from this, living right on the interface, reacting the reduced ions in the crust with oxidised ones from above. Photosynthesis has acted to steepen this energy gradient, as highly reduced organic substances collect as sediment over the land and ocean-floor and oxygen is pumped into the air and water. Over evolutionary time, the potential difference of the entire planet has increased.

Many of these redox reactions occur naturally, so the only way life can use them to extract energy is to run them more quickly than they would normally proceed. Once the chemicals reach equilibrium there is no energy gradient and life is extinguished. This is why enzymes are so crucial; they allow life to outrun the geological processes that would eventually remove the gradient.

For example, the rusting reaction of ferrous iron oxidation normally takes years, so enzymes can easily improve on this and keep ahead. Some reactions do not occur at all at low temperatures without the action of life, while others, such as the one between iron oxide and hydrogen sulphide, proceed so quickly that enzymes cannot perform any better.

This need to accelerate abiotic reactions leaves a very distinctive signature in the environment. If you were to descend deep below the surface of, for example, the Black Sea and look at the concentrations of different chemicals as you went, you would see a series of sharp gradients and peaks. Oxygen is present at high concentrations near the surface but rapidly disappears, due to aerobic respiration. Lower down are peaks of reduced nitrogen and iron, caused by chemoautotrophs and deeper still are rising levels of other reduced chemicals, such as ammonia and hydrogen sulphide. Without life, the reactions producing these chemicals would run far too slowly ever to set up gradients like this: diffusion and water currents would mix them more quickly than they could form. The fact that the Black Sea is not an even mix of chemicals is an iron-clad proof of biological processes. At different scales, all ecosystems show such tiers: across metres in lake water, centimetres in sediments and micrometres in biofilms (such as on the surface of stromatolites). This gives astrobiologists the ability to detect alien ecosystems without making any assumptions about the appearance, genetics or metabolic networks of the organisms within them. Life betrays its presence in a layered series of redox zones in the environment, a clear indication of the driving force of biologically-accelerated chemistry.

Is there another way?

In the first part of this chapter, we took a whistle-stop tour of how terrestrial life works by storing genetic information and generating energy to maintain its complexity. But what makes us think that life on another world would be anything like life on this one? First, many of the building blocks used by terrestrial life, the sugars, amino acids and bases, seem to be extra-terrestrial. They are produced in the great clouds of dust between the stars and during the formation of the solar system, as you will see in more detail in Chapter 3. But could these subunits, or others slightly different from the ones terrestrial life chose, be put together in other ways to create living systems? Since there is great metabolic diversity within terrestrial life but only a single genetic system, we'll look at RNA. Is the particular make-up of terrestrial RNA, and the closely-related DNA, unique in any way or could similar molecules perform an adequate job? After that, we'll step even further back and question what might be so special about carbon-based chemistry or the use of water as a solvent. Finding carbon and water-based life is the assumption which drives our current exploration of the solar system but how solid is the reasoning? But first, a closer look at RNA's molecular cousins.

DNA/RNA

During the first stage of the emergence of life, long before cells, a time when only replicating molecules inhabited the Earth, prebiotic chemistry would have produced a wide variety of different sugars and bases. So why does all current life on Earth use only two closely-related polymers to store its genetic information? What's so special about the particular subunits used in RNA and DNA? Are they superior to the alternatives, or was their survival more of a fluke, an evolutionary happenstance? The prevalence of RNA today may well be an accident; that once the RNA system was randomly selected it was too difficult for evolution to go back. This is rather like the fact that computer keyboards retain the qwerty key layout of mechanical typewriters. This pattern is by no means optimal; it was designed to force slower typing and so reduce the number of jams. Despite a number of attempts, it has proved too difficult to change the qwerty layout and so this suboptimal system has become fixed.

To address the question of whether there are workable substitutes to RNA and DNA, researchers have been creating artificial alternatives. DNA is practically the same as RNA, differing only in an extra oxygen bonded to the ribose sugar ring and the use of the T base rather than the U. Both are composed of three parts: a sugar, a phosphate group which binds the sugars into a strand and the base. It is relatively easy to change the bases within a DNA strand; experiments have been performed to systematically substitute lab-made bases – changing the letters of the genetic alphabet. The structure of the DNA is not overly distorted, it still forms a double helix, despite the bonding between the base rungs being different. The reason why alien DNA probably would not use these non-standard alternatives is that they are less likely to be formed by the prebiotic processes thought to occur in interstellar gas clouds or on primordial planets.

Most experiments into alternatives focus on RNA, as is it thought to pre-date DNA in the evolution of life. Terrestrial RNA is built from a five-carbon sugar (the ribose): two main modifications to this have been tested in the lab; either the bonds linking neighbouring sugars are attached to different carbon atoms within the ring or a different sugar molecule is used. Our RNA is formed into a long strand by phosphate groups that bind the third and fifth carbon atoms of adjacent sugars; an alternative would be to link the second and fifth carbons. However, not only is the strand so formed much more vulnerable to snapping but the strength of pairing between bases is also severely weakened.

Ribose exists in different isomers (isomers are made of the same atoms but their configuration is slightly different). RNA made using some of these other isomers has been found to have very strong bonds between the base-pairs and so produces more stable helices than native RNA. The fact that an entire family of isomer-alternatives creates a stronger structure is very interesting in terms of the evolution of RNA, suggesting that natural RNA was not selected from the pool of alternatives solely on the basis of its stability. Perhaps a moderate degree of base-pairing strength is better, because it allows easier separation of the two strands during replication. Furthermore, helices with stronger base-pairing tend to make more mistakes, binding sequences that don't match perfectly.

The final major modification that can be made to RNA is to change the type of sugar used. Ribose is a five-carbon sugar; what about swapping it for one with four or six carbons? When the experiments were done, it was realised that six-carbon sugars, such as glucose, don't make very good information storage polymers. The bonding between the base pairs was woefully inadequate; not only was it very weak but the bases were much less particular about which they paired with – A would happily bond with G, rather than only with T. The larger sugar units are probably just too bulky and interfere with close bonding. The results from four-carbon sugars were much more exciting. The polymer TNA, made using the sugar *threose*, allows base pairing of a similar stability and accuracy to RNA and forms into a perfect double helix. TNA can even cross-pair with RNA and DNA, something that was impossible for most of the other alternatives. This has raised a lot of interest, due to the simpler nature of TNA, based as it is on a four-carbon sugar. Such a sugar would form more easily in prebiotic reactions and so TNA is much more likely to have arisen on the early Earth. The importance of this will become obvious in Chapter 4, when we look at the evolution of replicating molecules and the difficulties of explaining how DNA, or even RNA, ever appeared in the first place.

Experiments into worthy competitors to RNA have, so far, gone up a lot of dead ends. Many alternative polymers cannot form stable double helices or are inferior to RNA in other ways. The feeling at the moment is that alien life which evolved from a soup of prebiotic chemicals similar to that on the early Earth is likely to use either RNA, or a very similar molecule, to store its genetic information. However, the landscape of all possible alternatives to the sugar, bond-placement and base of RNA/DNA is truly vast and has barely been explored, let alone for a completely different kind of polymer altogether. The next step is to test which polymers are able to replicate and develop over time under Darwinian evolution.

Carbon

Polymeric chemistry is absolutely essential for our kind of life, which uses the nucleic acid polymers, DNA and RNA, to store and transmit information and carbohydrate polymers, like starches, as an energy stockpile. The amino acid polymers, proteins, are the most diverse bunch of all. Carbon forms the backbone of all of these molecules; without it, life on Earth would quite simply be impossible. It is hard to imagine any other element with such a propensity for building molecules with the complexity necessary for life.

Silicon is the most frequently mentioned alternative. It is situated directly below carbon on the Periodic Table of the elements, meaning that it behaves very similarly to carbon in chemical reactions – for example, it too is able to form four bonds at a time. The silicon atom is larger than the carbon atom, meaning that less is formed in the cores of stars and so there is much more carbon than silicon in the Universe. However, due to the process that created the solar system, silicon is much more abundant on Earth, almost thirty per cent of the crust, so silicon is unlikely ever to be in short supply on a rocky planet. Considering this, if complex silicon chemistry were possible on Earth it ought to have resulted in life based on silicon, rather than its rarer chemical cousin, carbon. Silicon is used by terrestrial life, such as in the protective shells built by diatomic algae, it just doesn't form the basis of our polymeric or metabolic chemistry.

The fact that silicon is a larger atom is important in another way. The bonds it makes with other atoms are generally weaker than those carbon makes, producing very fragile polymers, at least under the surface conditions that have been prevalent throughout Earth's history. Under very different conditions, such as high pressures and high or low temperatures, silicon may indeed form sturdy enough polymers. However, we cannot even begin to guess what such a biochemistry would be like, making designing tests to look for it impossible.

Another difficulty of silicon is the compound it forms with oxygen. Carbon polymers can be oxidised, such as during respiration of carbohydrates, to release carbon dioxide gas. The analogous process would produce silicon dioxide, which is sand. This is a hard, insoluble, solid; very tricky for life to deal with. The final reason to favour carbon-based life, at least in our corner of the galaxy, is that the organic building blocks are rife throughout the heavens. Primordial planets are thought to have been utterly smothered with falling vital carbon compounds early in their development. Similar silicon compounds are not seen in space and so it really does seem that searching for carbon-based life is the only sensible option.

Water

Whether extra-terrestrial life would be dependent on liquid water is much less definite. Water has an impressive list of skills. First, liquid water is an exceptionally good solvent. The two hydrogen ends of its V-shape are slightly positively charged, whereas the oxygen atom at the point is slightly negative. The water molecule is *polar* – it has both positively and negatively charged ends. This means water molecules can form weak 'hydrogen bonds' with each other, as well as with other polar substances or ions, to dissolve them. Having molecules available in a dissolved state is crucial for the rapid chemical reactions of life; it also helps in gathering essential nutrients and expelling waste products. Organic life in a solid state is impossible, because molecules are caged up and unable to move and react; and a gaseous life form would have the opposite difficulty of molecules rapidly moving away from each other and never reaching concentrations high enough for fast reactions.

The hydrogen bonds in water ensure it is liquid, and so available for biology, over a wide range of temperatures (between 0 °C and 100 °C at Earth's sea-level pressure – see Figure 5 on page 61). Water is also quirky in that it expands when it freezes. This results in ice being less dense than liquid water, so that it floats on top of the liquid water and insulates it, preventing the entire volume from freezing solid. Water can also absorb a lot of heat energy without a big rise in temperature, which could damage biological molecules. This makes it an excellent thermal buffer, protecting cells from wild temperature swings and preventing disruption of vital activities like enzyme action. Water is also very chemically useful, contributing either oxygen or hydrogen atoms across a range of biochemistry, such as the hydration reactions that break down proteins or starch polymers into their subunits.

No other known liquid combines all these amazing properties but we don't know which of them actually were relevant to the emergence of life. Could another solvent, even one that possesses fewer of these attributes, still be suitable for biochemistry? For example, it is often stated that the polar nature of water is essential for the correct folding up and functioning of proteins. But molecules called peptoids, chemically very similar to proteins, can fold up in pure methanol. We may even be asking the question the wrong way round. The properties of water seem fortuitously fine-tuned for life but it is equally possible that life on Earth is fine-tuned for a watery environment. Terrestrial life evolved to exploit the specific properties of water simply because it is the only solvent available in large quantities on our planet. Terrestrial life is well-suited to water, rather than visa versa, and there may be no good chemical reason why life couldn't emerge in other solvents. In fact, water is, in some ways, inappropriate for life: it is fairly reactive and tends to break apart complex organic molecules. RNA, for example, is particularly short-lived in water.

What alternatives are there? Pure ammonia is liquid between −78 °C and −33 °C at Earth's atmospheric pressure (a range of forty-five degrees − half that of water) and can form hydrogen bonds to dissolve many organic substances. Ammonia is a very common compound in the galaxy, present in the dust clouds in outer space and as liquid droplets in the clouds of Jupiter. A mixture of water and ammonia can exist as a liquid at temperatures much below that of pure water and represents a hybrid solvent. This is particularly relevant, since there may be aquifers of water and ammonia beneath the surface of Saturn's moon, Titan. Another chemical, formamide, is liquid across a wide range of temperatures and pressures, dissolves salts and has other properties similar to water. Other possible solvents include methane (liquid on Earth at around −160 °C) and liquid nitrogen (fluid at −196 °C). However, although they could in principle support an organic biochemistry, the much lower temperatures are a big issue. Even if the solvent is liquid in such cold climates, the chemistry will not be the same. The colder the conditions, the more slowly processes occur and the more effective an enzyme must be to get reactions to proceed at a useful rate. At these temperatures, biochemical reactions might proceed so sluggishly as to render life impossible.

As with non-carbon-based life, a non-water-based ecology is so alien as to make it difficult to design experiments or instruments to send to other worlds. We haven't even begun to engineer an alternative metabolic network that operates within water, let alone one based on an entirely different solvent with chemical properties we've barely researched. NASA's astrobiological motto is 'Follow the water' because, on Earth at least, where there's liquid water there's life. This is likely to remain the focus of astrobiology research for the time being − hunting in conditions we know can support life, rather than speculating about other, perhaps fantastical, possibilities.

What do we know about life? We know it must exist on an energy gradient, extracting energy from the environment to push its own complexity uphill. Life on Earth is powered by two fundamental methods of energy extraction: electrons liberated either from photosynthesis or redox reactions are used to generate a proton gradient that runs ATP-synthase or fermentation of reduced organics directly produces limited amounts of ATP. Life must also contain a coded self-description and replicate this information for subsequent generations. These requirements are satisfied in terrestrial life by polymeric chemistry, which needs organic substances (carbon) dissolved in a fluid medium (water). Other solvents are feasible but the versatility of carbon is thought to make it a universal of chemical life. Exactly which polymers extra-terrestrial life might build with carbon is much less clear. DNA/RNA and proteins are universal on Earth but we've only just begun to explore the alternatives in the lab.

2

Extremophiles

The spark that really ignited interest in the possibility of life on other worlds was the discovery of so-called *extremophile* life on Earth – organisms thriving in hostile environments, previously assumed to be absolutely sterile. Interest in this branch of microbiology has grown enormously since the 1990s; now it feels as if in any location we care to check we invariably find life flourishing. Life, once started, seems to have become utterly irrepressible, adapting to fill almost any damp niche. There is evidence for liquid water on many planets and moons within the solar system; indeed it ought to be abundant on any planet at an appropriate distance from its star, so these discoveries at the extremes of our own world are obviously tremendously important.

The envelope of life

There are three main parameters that affect the functioning of cellular biology: temperature, acidity and salinity; different organisms can survive over different ranges of these conditions. Taking all cells, these ranges of tolerances can be combined and plotted on a three-dimensional graph to show the total scope of life on Earth. This describes the survival envelope of life as we know it, from acidic cold water in one corner to alkaline hot brine in the opposite. The exciting realisation is that parts of this bulgy cloud overlap with the conditions we believe exist on other worlds. In terms of temperature, acidity and salinity, certain terrestrial life would be perfectly content in extra-terrestrial settings.

We'll take each of these in turn, explaining not only what kind of environments such extremophiles cope with but also a little about the evolutionary adaptations that ensure their survival. The complexity of the eukaryotic cell means it is sensitive to physical and chemical extremes, so much of the outer regions of the survival envelope are populated only by prokaryotes. However, we should remember that extremophiles are extreme only in relation to the capabilities of other cells. There is no such thing as a 'standard' environment on Earth and we don't have much of an idea about the conditions under which life first arose. Extremity is in the eye of the beholder – from the point of view of a bacterium dwelling in the crush and heat of the deep underground, the cells eking out a living in the bitter cold and low-pressure environment of the surface, toxic with corrosive oxygen and awash with UV radiation, are the evolutionary freaks.

Temperature

The absolute limits of temperature survivability are set by the freezing of water on one hand and its boiling into vapour on the other: active life requires liquid water. But there are other factors that modify water's freezing and boiling points away from the 0 °C and 100 °C that we are familiar with. Pockets of briny water within solid ice can remain liquid at −20 °C and water in the deepest parts of the ocean only starts boiling at 400 °C (similarly to a pressure cooker).

Surviving very low temperatures is often not a problem. Many organisms can be frozen and later re-animated – fertility clinics regularly store eggs in liquid nitrogen at around −200 °C, for example. However, important factors limit how cold a cell can be yet remain active and grow. Such cold-loving bacteria are called *psychrophiles*, as shown in the table below.

Parameter	Extreme	
	low	high
Temperature	*psychro-*	*thermo-*
pH	*acido-*	*alkalo-*
Salinity		*halo-*
Pressure		*baro-*

The main problem that must be overcome at low temperatures is that the cell's proteins, particularly the enzymes, become very rigid. The catalytic function of enzymes depends on them being able to change shape but the colder it is, the less thermal vibration helps them flex. Cells adapted to near-freezing temperatures have evolved enzymes with much looser bonds, which means they remain supple, even at very low temperatures, and the cell can keep its metabolism going. For similar reasons, their cell membranes are also kept much more mobile. The danger facing such cells is that if the temperature rises too far above their optimum, these looser proteins denature, just like the frying egg. Cold-adapted bacteria effectively begin to cook at just 20 °C and if the mis-shapen proteins don't kill the organism, the spilling out of the cell contents as its membrane ruptures certainly will.

Psychrophiles are at home in Antarctic waters and deep-sea sediments; certain psychrophiles are active within Antarctic ice at −20 °C. As the sea-water freezes, the salt becomes increasingly concentrated in small pockets, preventing the water within them solidifying, producing a labyrinth of interconnected channels and pores. Such low temperatures result in very slow reproduction times of up to six months (compared with bacteria living within the human gut which divide every twenty minutes). Much of the Earth's surface biosphere is cold – more than ninety per cent of ocean water is below 5 °C and so organisms adapted to the high pressures of deep water (*barophiles*) also tend to be psychrophilic.

Heat-loving organisms, *thermophiles*, face the opposite difficulties. They have evolved enzymes with extra bonds, to stop them shaking apart at high temperatures and membranes with reduced fluidity. Although higher temperatures allow chemical reactions to proceed much more quickly and so, in one sense, thermophiles have an advantage, above about 150 °C many organic molecules decompose. This sets an absolute upper limit on survivability and, so far, no organism has been isolated that can tolerate more than 121 °C (a record set by an archaeal chemoautotroph). The majority of extremely heat-tolerant organisms – *hyperthemophiles* – are archaea. Eukaryotes are comparative wimps; a few algae and fungi are able to cope up to about 60 °C and the Saharan Desert ant can forage under the midday sun at 55 °C. Chlorophyll degrades at 75 °C, so very little photosynthesis happens in hot pools. Another problem is that the solubility of gases such as oxygen and carbon dioxide decreases in hotter water and so many hyperthermophilic cells are anaerobic.

Thermophiles are normally found in environments like the flanks of volcanoes and around geysers and geothermal hot springs. These locations are usually rich with reduced chemicals from inside the Earth and so many thermophiles are chemoautotrophic, reacting hydrogen, ferrous iron or reduced sulphur compounds with electron acceptors like oxygen or nitrate. Extracting energy by oxidising sulphur compounds has one fairly nasty side-effect: these reactions produce sulphuric acid, often making the geothermal waters very acidic. The volcanic hot pools in Yellowstone National Park in the USA have sulphurous gases bubbling through them and can be extremely acidic. Bands of vivid colours are often seen along the edges of such pools, a reflection of the regions of minerals in various redox states created by different thermophilic organisms, the water getting gradually hotter, more reduced and less acidic as its depth increases. The thermophiles nearer the surface of the water must therefore also be adept at handling high acidity.

Acidity

Acidity is rated on the pH scale, which measures the concentration of H+ ions (protons) in the solution. A low pH signifies an acidic solution; a high pH an alkaline one (neutral is defined as pH 7). Concentrations of protons are a fundamental mechanism used by cells to transform energy; indeed, many of the organic molecules life is built of can lower pH, such as amino acids, nucleic acids and numerous small molecules involved in metabolism. Many *acidophiles* can tolerate as low as pH 2 – about the acidity of lemon juice. Some acidophilic species were first discovered when they spoiled tinned fruit; others are common in hydrothermal pools or in the water running out of metal mines where the ore is very rich in sulphur. The most extreme can survive at pH 0; a hundred times more acidic than lemon juice. Such high acidity, when combined with high temperatures, is extremely destructive to cellular molecules and so the most acidophilic organisms can only tolerate a relatively cool 60 °C.

The main difficulty acidophiles encounter is the stability of their proteins. The exact three-dimensional structure of a protein strand is held together by different kinds of bonds and these can be greatly disrupted by changes in the distribution of charges, such as an increase in acidity. Acidophiles protect their proteins by including more amino acids with neutral side-groups and by actively maintaining the inside of their cells as close to neutral as they can. They do this by pumping out protons as they leak back across the membrane or even by turning them to their advantage. For example, certain organisms oxidise iron outside the cell and transport the released electron inside to bond together oxygen and hydrogen ions and create water. Protons are thus removed from within the cell and those outside are allowed to flow inwards through ATP-synthase and generate energy.

At the other end of the spectrum, *alkaliphiles* survive in very alkaline waters, such as the soda lakes of the western USA and Lake Magadi in Kenya. These organisms tolerate conditions of up to pH 11 – equivalent to that of household ammonia. Under these circumstances, the concentrations of hydrogen ions are very low and cells have trouble using ATP-synthase to generate power and other essential ions, such as magnesium and calcium, precipitate out of the water as salts and so are available only at very low levels. The alkaliphiles cope by actively pumping in these ions and exporting others to maintain their interior at near-neutrality.

Salinity

A wide variety of organisms can thrive in essentially pure water but the other end of the scale seems to be much more problematic. Salt-loving *halophiles* must contend with a difficult osmotic balance (the difference in water concentration between the inside and outside of the cell). Osmosis is the process by which water tends to move into regions with highest concentration of solutes; that is, from most to least dilute. This means that cells living in a very salty environment are in danger of losing all their internal water to the outside. They combat this by pumping out sodium and keep their insides concentrated using benign chemicals, such as amino acids or glycerol. Thus they avoid the problem of having a very salty cytoplasm, which would disrupt and possibly denature proteins.

Hyper-saline environments are created by high evaporation rates from lakes, such as the Dead Sea, or by very salty water becoming trapped in ocean basins by less-dense water above. High salinity seems to be one of the less demanding physical constraints; even some higher organisms can survive the extremes. Ecosystems of brine shrimps feeding on dense algal blooms are often observed in brine lakes but as the water evaporates and the salinity approaches saturation point (where no more salt can dissolve) at thirty-five per cent dissolved salt – ten times more concentrated than sea water – only the halophilic archaea remain.

Although discussions on extremophiles often focus on only the most impressive instances in the bacteria and archaea, there are plenty of astounding examples among eukaryotes. These are very relevant to the possibility of complex ecosystems arising on alien worlds. True, eukaryotic cells cannot manage long periods at anything warmer than about 60 °C – the temperature of a cup of tea – but many higher organisms cope in freezing conditions. The wonderfully-named *grylloblattid* insects scavenge over the ice sheets on mountain tops for animals killed by the cold. They are more commonly called icebugs or rockcrawlers and are active well below freezing point (if you pick one up the warmth of your hand is enough to kill it). Due to the low temperatures, icebugs grow and develop exceeding slowly, taking up to seven years to complete a generation. The American skunk cabbage can grow in sub-zero temperatures, when the ground around it is frozen solid. It accomplishes this partly by using the proton gradient established by respiration not to make ATP but directly to generate heat to thaw the surrounding soil. Perhaps most remarkable of all complex eukaryotes are the *tardigrades* or water bears. These microscopic animals, less than a millimetre from head to tail, with eight stubby legs, superficially resemble arthropods such as lobsters but are deemed unique enough to deserve their own group. They are common on damp mossy patches almost anywhere from the tropics to the poles. When the going gets tough, tardigrades transform into a barrel-shaped 'tun' state and can hibernate like this indefinitely. As a tun, tardigrades can cope with pretty much any hazard thrown at them, including temperatures from −253 °C to 150 °C, intense X-rays, extremely high pressures or a vacuum. They are the true survivors of the animal world and a good model of the sort of complex life that might be found elsewhere.

Alien worlds on Earth

As well as identifying the limits of survivability on Earth – seeing just how far the envelope of life extends – astrobiologists are keen to study spots on our planet that are thought closely to resemble known extra-terrestrial environments. We'll look at three of these in detail: first, one thought to be a good analogy for life on the surface of Mars.

Antarctic dry valleys

Antarctica is not only the coldest place on Earth but also contains some of its driest regions. Its frigid air can hold little moisture anyway and as the winds blow across the southern continent, which is shaped like an upturned bowl, they are pushed to higher altitudes and dry out even more. These extremely desiccating winds blow along the top valleys, creating huge areas of totally barren, ice-free rock. There is less than half a centimetre of precipitation a year and even this meagre amount remains frozen for much of the time, inaccessible to life. During the long bleak winter the temperature in these dry valleys plummets to −40 °C, soaring to a balmy 1 °C in summer, a season that lasts but a fortnight. Animals that become stranded in these valleys inevitably die and are instantly freeze-dried. Some seals' bodies have been found that have been preserved for thousands of years.

When Robert Scott explored in 1903 he described them as the 'valleys of the dead' because he could see no signs of life. But if you look closely enough, you find living organisms, tiny populations of nematodes and tardigrades feeding on bacteria and detritus, thawed out for the all-too-brief summer. These organisms do not crawl over open ground but live in tiny crevices between the grains of boulders strewn across the desolation. Entire ecosystems of organisms are hidden within the rocks, giving them their name: *cryptoendoliths*.

Life is only possible because these cracks maintain a microclimate much more pleasant than the freezing air outside. The Sun's meagre heat warms up the rocks and the cracks trap the melted ice and protect their inhabitants from the brutal wind. For the photosynthetic lichen and bacteria which support these miniature ecosystems, the thin cracks provide another crucial service. Antarctica, due in part to the hole in the ozone layer, has one of the highest levels of UV on the planet, radiation that inhibits chlorophyll and damages proteins and DNA. Being buried beneath just a few millimetres of rock means that much of this lethal UV light is blocked but enough visible light still gets through to drive photosynthesis – the translucent stone acting as a very effective filter.

Such endolithic habitats are thought by some to be possible on the surface of Mars. Mars' ambient temperature, even on its equator, is extremely cold for most of the year but cracks in rocks could absorb enough warmth to collect pools of meltwater and support life for short periods over the Martian summer. Mars, without a thick atmosphere or ozone shield, is totally unprotected from the UV radiation – the DNA-damaging effect is four hundred times greater than on Earth. Absorption of the UV light by a thin layer of translucent rock could provide similar filtering effects as in the dry valleys and so allow the continuation of photosynthesis. However, there are still some pretty enormous problems facing Martian surface life: first, liquid water is not generally possible under the current atmospheric conditions – Mars makes the dry valleys seem hot and humid – and second, the solar UV radiation is thought to have rendered the exposed Martian ground very oxidising, tearing apart any organic molecules that may have accumulated. It is hard to imagine even endolithic habitats providing the basic requirements for life but Mars has not always been this unpleasant; in Chapter 5, I shall discuss the possibility of surface life in ancient Martian times and the chance that it thrives, even today, deep underground.

One mechanism for creating crevices in rocks is likely to be widespread in solar systems throughout the galaxy. Meteorite and cometary strikes are commonly thought to be agents of destruction but the immense shock waves generated by such impacts fracture nearby rocks, opening up an enormous number of potential endolithic habitats. For example, beneath the thick ice of Devon Island, the largest uninhabited island on Earth, in the far north of the Canadian Arctic, lies the Haughton crater. This twenty-four kilometre wide pit was excavated about twenty million years ago by an impact and is now buried beneath a thick sheet of ice. A ring of shocked volcanic rocks around the crater has been found to be twenty-five times more porous than rocks elsewhere. These cracks are inhabited by photosynthetic organisms, thriving for a few months a year within their clement micro-environment. The Martian surface is littered with rubble broken apart by impacts and so its surface is not short of similar endolithic habitats. Equally important is the possibility that the first cells to colonise the land masses on Earth made use of these prime residential opportunities. The rate of impacts during the early solar system was thousands of times higher than today and the rocks of the first continents were probably extensively fractured.

There is another habitat resulting from a meteorite hit that may turn out to have enormous astrobiological significance in freezing environments. Not only does an impact create a wide area of shocked rocks but also a focal point of immense heat, coming both from the energy of the strike itself and through deeper, hotter, crust being brought up to the surface, melting the surrounding ice into a geothermal lake. Bacteria rising up from below or arriving on the wind quickly colonise this oasis of warm water in a land of permafrost. It has been calculated, for example, that the Haughton crater bore a pool of water at temperatures exceeding 50 °C for many thousands of years. At the tail end of the heavy bombardment, the larger impacts would have created surface hydrothermal environments capable of persisting for many millions of years, time enough to be significant in the evolution of life. Lake water percolates into the hot crust beneath and seeps back out again rich in the dissolved minerals which can support colonies of chemoautotrophs (like the hydrothermal vents we'll visit in the next section). Over time, the stored heat gradually escapes and the lake freezes over but the water beneath this insulating cap remains liquid for many more years. The potential for life decreases as the energy gradient it depends on dissipates and the ecosystem winds down to a final collapse.

The southern hemisphere of Mars is very heavily cratered and thought to contain a great deal of subsurface water. Such impact hydrothermal systems might therefore have been very important for astrobiology on the primordial Red Planet, both in the emergence of life and in providing new niches for subsequent colonisation. Intriguingly, craters have been spotted in the polar regions of Mars which still contain lakes of frozen water. It is not yet known whether these would ever have been geothermal pools capable of supporting life but the possibility is enticing. As the lake water froze it could have preserved cells within it. The hope is that future probes will dig samples from such ice and thaw out any trapped inhabitants in an attempt to revive them.

Deep sea vents

In the late 1970s, when the Voyager space probes were finding erupting volcanoes, new moons and rings in the Jupiter system, something equally extraordinary was discovered in the dark depths of Earth's oceans. Researchers working in submersibles were familiar with barophilic life that coped with the crushing pressures of the ocean bed. Barophiles had even been discovered at the bottom of the Mariana Trench, the deepest point of the Earth's crust, where the pressure is over a thousand times greater than at sea level. Although photosynthesis is impossible in these pitch-dark depths, animals are at home on the deep abyssal plains, subsisting on the organic detritus which snows down from the productive, sun-lit, regions far above. Great herds of sea cucumbers (a translucent tube-shaped animal, related to the starfish) are occasionally seen travelling across the ocean floor, churning up clouds of sediment like stampedes of buffalo on the savannah.

However, at these depths, food is scarce and this, combined with the low temperatures, results in low populations of animals and very slow growth rates. This was thought to hold true for the whole ocean bed until, in 1977, researchers stumbled across the most astonishing exception; a discovery that revolutionised biology.

In the mid–Pacific, a long scar runs across the face of the planet, a spreading centre where fresh magma pushes up to form new oceanic crust and forces apart two tectonic plates. Along this ridge, two and a half kilometres beneath the waves around the Galapagos Islands, scientists crammed into a tiny submersible discovered teeming oases of life in the cold dark desert. There, water seeps through the cracked crust and comes into contact with the hot magma chamber just beneath the ridge, becoming super-heated

Figure 4 The dense plume of reduced inorganic ions spewing out of a mid-Pacific black smoker and the mysterious mouth-less tube worms and other life huddled around such vents.

and leaching minerals out of the surrounding rock, particularly iron and sulphur compounds, until it becomes saturated. The water is forced back up, gushing out of localised spots – *hydrothermal vents* – on the ocean floor. The emerging water can approach 350 °C and is prevented from boiling only by the crushing pressure of the deep water, but it cools rapidly as it mixes with the frigid ocean around it. Its carrying capacity drops and many of the dissolved minerals instantly precipitate out. This produces a great cloudy plume of dark particles, which gives these vents their nickname 'black smokers'. Some of this mineral material accumulates around the vent, building up tall chimneys that funnel the hot water out at the top, as shown in Figure 4. These occasionally collapse under their own weight but some grow to truly monstrous sizes, such as the 'Godzilla' black smoker, off the coast of Washington state, that is higher than a sixteen-storey skyscraper.

It wasn't the geology of the area that surprised the scientists so much as the wealth of life thriving there. Pale crabs and shrimp scuttled around the chimneys in the cooler water, feeding off giant, dinner plate-sized clams (molluscs estimated to grow three hundred times more quickly than their relatives on the abyssal plains – a testament to the abundance of energy pouring out of the vents). Most astounding were the tube worms, alien-looking animals with blood-red heads, up to one and a half metres long, and completely new to science. Such vent ecosystems are extraordinarily ancient. As we will explore in Chapter 4, it may be that life arose around hydrothermal vents on the primordial Earth. The mystery of the tube worms and clams only deepened when some were brought to the surface for closer inspection. They were found to be completely lacking either mouths or digestive systems. How could they feed to stay alive? Their secret was not discovered for another four years, and lies in symbiosis, a particularly close association between two different organisms.

The vent jet is laden with reduced ions brought up from within the crust, which support a large population of chemoautotrophs around the edges of the plume. These organisms must also be thermophilic (the most heat-tolerant organism on Earth lives around these vents, coping with temperatures of up to 121 °C). Many of these chemoautotrophs catalyse the oxidation of reduced sulphur or iron compounds, using the redox reaction to provide energy and to fix the copious carbon dioxide present into required organic molecules. The tube worms survive on these chemoautotrophs, not eating them but hosting a large colony within a specialised organ almost half the size of the worm. The red head of the worm functions like a gill, stuffed full of haemoglobin proteins (very similar to the pigment in human blood but independently evolved) which absorb oxygen and hydrogen sulphide from the water. The chemicals are passed to the chemoautotrophs, which perform the redox reaction and supply nutrients to the worm, in return for this shelter and food supply.

There was an even bigger shock in store for the scientists. Less sunlight penetrates these ocean depths than reaches Neptune and yet they found shrimp with light-sensitive organs on their backs. What on earth could these animals be seeing? The most likely explanation is that the shrimp sense the faint light given off by the superheated vent water, thus avoiding getting scalded. But this realisation presents another possibility. If animals make use of this dim glow, could cells photosynthesise by it? In 2005, photosynthetic bacteria were indeed collected in vent water samples. The proof is still lacking but if organisms can survive on geothermal light alone this opens up a completely unexpected energy source for astrobiology.

Similar vents have now been found on sea-floor spreading ridges all over the world, although only those in the Pacific host giant tube worms and clams. Each vent has a particular chemical composition, water temperature and flow rate. Some mix with the surrounding sea water very gently, so don't produce concentrated plumes of precipitation. In terms of possibilities for astrobiology, hydrothermal vents may represent an almost universal mode of life. Thermophilic chemoautotrophs can support an entire ecosystem of bacteria and animals, on inorganic redox reactions alone, far from the light of the Sun. However, even this deep life depends indirectly on photosynthesis, as much of the dissolved oxidising power in the sea derives from plants in the sunlit waters above. Black smoker life can run entirely on abiotic oxidants, as indeed the first cells are believed to have done, but the lower energy availability would severely limit the size of the ecosystem. Any planet or moon with a watery ocean and source of internal heat could provide such a habitat, with subsurface ecosystems at the interface between the hot reduced crust and the cold oxidised water. We'll explore this possibility further when we consider the case of Europa, the ice moon of Jupiter.

Deep basalt aquifers

Hydrothermal vents are a promising model for extra-terrestrial life but on Earth they are still, partly, supported by photosynthesis. However, there are a few examples of life that is absolutely isolated, completely independent of sunlight and self-reliant in energy and organic substances. These ecosystems, powered by chemoautotrophs, provide the best general model on Earth for life elsewhere.

One is in a thick layer of rock, lying beneath what is now the Columbia River basin in Washington state, USA. This basaltic rock was formed by a series of large floods of volcanic lava across the surface around ten million years ago. Basaltic rock is rich in reduced iron and reacts slowly with aquifer water to release hydrogen. Chemoautotrophic arcahaea living a kilometre or so deep within it derive all their energy by oxidising this hydrogen, passing the freed electron to dissolved carbon dioxide and so fixing it. The products of this redox reaction are water and methane gas, so these organisms are called *methanogens*. These anaerobic methanogens and the heterotrophs that feed off them are believed be completely independent of oxygen or organic molecules produced by photosynthesis above and so are called *Subsurface Lithotrophic Microbial Ecosystems* (SLiMEs). A further advantage for these communities is that they do not require hotspots of active geothermal heating like the black smokers, just a slab of reduced volcanic rock.

Just how deep life extends into Earth's crust is another topic of intense research. Miles underground, there are still minute pores and fractures within the rock, filled with water and amply big enough for bacteria. The furthest scientists have ever poked into the Earth's crust and sampled is 5.3 km down a bore hole in Sweden. Even at the bottom, at a depth where the temperature of the rock is more than 70 °C, there is a variety of active heterotrophic bacteria. It is likely that there is no survival limit to depth, so long as the ambient temperature does not become lethal. The highest temperature known to be tolerated by a thermophile is 121 °C. This point is reached at different depths depending on the local thickness of crust but can be as much as 10 km down in some sedimentary rock formations.

The cells of a deep SLiME are extremely slow growing and live at low density, at some sites only a few thousand cells per gramme of sediment, about a millionth that of a fertile topsoil. But considering the vast volume of the upper crust on Earth, this could still constitute a staggering mass of living material. The 'deep hot biosphere' has been estimated, by some researchers, to exceed the sum total of all surface life. In many ways, these subterranean fissures are ideal habitats – their unchanging environment provides a steady, if slightly limiting, flow of chemical energy, with no UV or energetic cosmic radiation and the cells are protected from all but the most devastating catastrophes above. The deep hot biosphere would be unfazed by even a large impact or global glaciation.

Basaltic rock, water and dissolved carbon dioxide are thought to be common substances on any terrestrial planets or moons capable of volcanism and so many believe the SLiME model to be a sure-fire method of supporting alien life. (This assumes that life has evolved: it is not yet clear whether life could emerge in such an environment rather than just colonising it.) Such conditions are believed to exist far underground on Mars. The top layer of Martian rock has been heavily broken up by impacts; its subsurface is thought to be loaded with ice-filled pores. This permafrost would melt around geothermal hotspots and react with the basaltic rock. On Earth, archaea that metabolise the hydrogen release methane: intriguing plumes of methane have recently been discovered on Mars. Could this be the signature of a deep SLiME layer on the Red Planet? We'll consider this possibility in Chapter 5.

Earth's biosphere is a vanishingly thin film across the very surface, sandwiched between the rock-melting heat of the mantle below and the cold void of space above. We have found cells roughly 5 km deep into the crust and around 40 km up into the thinning atmosphere. But even within the habitats of just this one planet, organisms cope with a staggering range of hazardous conditions. Extremophiles thrive in boiling acid, freezing brines and deep underground living off naught but gases bubbling from the rocks. But how extreme is *extreme*? How hostile an environment can life tolerate and continue growing? Could life cope even with exposure to outer space? Let's look at this possibility now and see how cells could be transferred between the planets.

Panspermia

The theory of panspermia, meaning literally 'seeds everywhere', was first widely discussed in the early 1900s. It was thought that hardy microscopic cells, such as bacterial spores, could be dispersed throughout the galaxy, blown by the radiation pressure of their star. The original theory fell out of favour, since unprotected cells would never survive the aeons of time necessary to voyage between the stars. But the essence of the idea, adapted with the notion that cells could be transported between planets and moons aboard meteorites, has recently been resurrected and is rapidly gathering encouraging evidence.

There are essentially three hurdles that bacteria must leap to survive the journey between two planets: ejection from their home world within a chunk of their surrounding rock, exposure to the hostile space environment during transit inside this meteorite and re-entry through the atmosphere of and impact on to the destination planet. Calculations and experiments carried out on each of these stages generally show no acute problem with the prospect of panspermia.

It was originally thought that living cells would be unlikely to survive being launched from a planet. An impact would need to be very energetic to throw up fragments of the target surface at sufficient speed to punch out through the atmosphere, escape the gravitational pull of the world entirely and enter interplanetary space. Such a violent strike would dump an enormous amount of heat into the ground, flash-boil any water and send an immense pressure pulse rippling through the solid rock, shaking and crushing all nearby life. Even the acceleration needed to be thrown clear of the planet ought to be enough to squash any cells flat. This is certainly true for the ground closest to the impact point but mathematical analysis has found an interesting surrounding region, the *spallation zone*. The pressure pulse propagating down through the crust is refracted back up to the surface, where it meets a more direct wave heading sideways away from the impact site. The two merge, and in certain regions destructively interfere, cancelling out much of the intense pressure. You can see exactly the same process when you throw a stone into a pond. In some places two crests coincide and combine together, in others a crest and a trough meet to produce flat water. Within the spallation zone, lumps of rock are thrown upwards at high velocity without being subjected to immense shock. This process is particularly efficient if the impacting body ploughs into the ground at a glancing angle – much of the rock above the strike explodes outwards relatively unharmed yet a single impact can hurl millions of tonnes of surface rock into space. The very brief pulse of heat means that it does not travel very far into the boulders and so bacteria insulated in the centre are safe from being cooked. Several lumps of Martian rock have been discovered on Earth, blasted off their home by just this process. When analysed, they were found to have been subjected to relatively minor peak pressures and their interiors not heated above 100 °C. Lab experiments have addressed the question of whether living cells could actually survive these conditions. Porous pellets of ceramic have been soaked with bacteria and then fired into a hard target at extremely high speeds, up to 12,000 miles per hour. During the crushing impact, the cells are subject to huge forces of deceleration but between one in ten thousand and one in a million bacteria recover and continue growing. This shows that a significant number of the inhabitants of surface rocks could survive the traumatic experience of being forcibly launched into space.

Once off their home world, things get decidedly worse for the bacterial stowaways. Space is an extremely hostile environment, about as far removed from any terrestrial niche as could be imagined. It is extremely cold, a near-perfect vacuum, bathed in radiation and in free-fall. Weightlessness does not appear to be a major problem for the functioning of a cell and many organisms survive the deep-freeze without harm. However, radiation is likely to be a major limiting factor: UV can kill an unprotected cell within minutes, a hazard exacerbated by the extremely desiccating effect of vacuum. However, a filter of just a few millimetres of rock is sufficient protection, as for the endoliths in Antarctica. The background radiation of highly energetic particles surging out from solar flares and supernovae throughout the galaxy is much more of an issue. The Earth's magnetic field and thick atmosphere shelters surface life from this radiation but bacteria within a meteorite would need several metres of solid rock for complete shielding. Cascades of these energetic particles would steadily degrade the delicate molecules in a dormant cell, imposing a time limit within which the meteorite must make its landfall on another planet before all the cells are killed. The extreme desiccation caused by the vacuum of space will also damage their molecular structure. The carbohydrates, proteins and DNA within a cell begin irreversibly to cross-link, undergoing the same chemical reactions as a steak browning on the grill. All these factors have been tested on live cells launched into orbit and exposed to the space environment. As expected, cells offered no protection against UV radiation died almost instantly but appreciable numbers survived the six-year mission with only minimal shielding, even if only that provided by the few layers of dead cells above.

Assuming that a population of cells is sufficiently protected from the radiation environment, could they remain in suspended animation for long enough before their meteorite crash-lands on another planet? To transfer between Mars and Earth, it would typically take fifteen million years for a meteorite's orbit gradually to distort until it begins crossing the Earth's path. Some simulations show a tiny fraction of rocks are ejected on a much more direct trajectory, being delivered 'first class', within just several thousand years, but in general, for panspermia to be plausible bacteria must be able to survive in space for many millennia.

Lying dormant turns out to be something that bacteria are rather good at. Many bacteria, such as those common in soil, form resilient spores when conditions start turning too cold or too dry or food runs out. This desiccated capsule is protected by a hard outer casing and the DNA within stabilised by special proteins. Such spores are very resistant to environmental assaults, including radiation, vacuum, temperature extremes and hazardous chemicals and they simply reawaken once their surroundings become favourable again. Other prokaryotes and even some eukaryotes, like the tardigrade, can wait long periods in a dormant state where growth and metabolism are paused. Just how long some cells have been dormant yet remained viable is astounding. Dry spores have been revived after 3,500 years in Egyptian tombs, psychrophiles after being frozen in ice for around 500,000 years and bacteria from forty million year old amber revived. Possibly the longest sleepers of all, however, is a sample of halophiles isolated from a large salt deposit deep beneath what is now the North Sea. These deposits were created by a lake which evaporated during the time when Europe was positioned over the equator. Bacteria have been extracted from the ancient fluid trapped within the crystals and seen to re-awake. If the cells are as old as the rock, a hotly debated claim, they would have been dormant for a staggering quarter of a billion years. All in all, dormant cells or spores buried right in the core of a reasonably-sized chunk of rock are indeed likely to survive in space for extended periods.

The final obstacle is the actual arrival. In many ways, the hazards of safe delivery to a new planet mirror those of departure. If the planet has a thick atmosphere, the surface of the plummeting rock will be substantially heated by air friction, producing a shooting star. But the larger boulders that have already provided radiation shielding for their cargo of cells will also protect them against high temperatures for the few seconds of re-entry; the bulk of the meteorite safely insulates the bacteria deep inside, even if the outside is melted into a charred crust. To test this, bacteria-soaked pieces of volcanic and sedimentary rocks were been embedded in the heat shield of a satellite. The rocks survived largely intact but unfortunately, the biological samples had fallen off and so could not be analysed for survivors. Another attempt is planned soon.

Planet-fall is a sharp deceleration, which sends a shock wave racing through the meteorite as it hits the ground. We've already seen that an appreciable fraction of cells survive such experiments, and particularly if the meteorite drops into ice or the ocean. Smaller grains of rock might provide only limited long-term shelter from radiation but, on the other hand, would tumble through the atmosphere without significant heating and land softly on the ground.

Calculations show that the exchange of rocks between planets is actually quite common. For example, since the formation of the solar system over a billion fragments smaller than a few metres across have been blasted off Mars without excessive heating or shock. About five per cent of these meteorites arrived on Earth within eight million years. Some forty of these interplanetary immigrants have been found; I shall tell the story of a particularly notorious one, with the rather bland name of ALH84001, in the chapter on Mars. The reverse traffic, Earth-Mars, would have been about a hundred times less, due to the stronger gravity of our planet. If life did emerge very early in the solar system on Mars or perhaps Venus, as well as on Earth, there would have been a busy cross-fertilisation of cells as the inner planets were battered by the high rate of impacts. The survival rate of bacteria within meteorites that land on ice rather than rock is much higher, so terrestrial colonisation of the moons of the gas giants is worth considering – although throwing a rock that far into the outer solar system is tricky; exchanging life in this direction is perhaps a million times less likely.

All three stages of panspermia, ejection, transfer and arrival, offer very low survival rates for the organisms aboard the meteorite. Each one has perhaps only a one in a hundred million chance of ending the voyage alive, odds more commonly encountered in lottery jackpots. But every lump of rock ejected from its home world could hold billions and billions of tickets to this sweepstake. Lithoautotrophic cells have been found living in the pores of deep basalt rocks at densities of up to one hundred million per kilogram and the abundance of spores in surface rock is a thousand times higher again. Even a small chunk of

Earth's crust ejected into space would be literally teeming with life. And it takes only a single bacterium to survive the interplanetary voyage, to reawaken, to grow, to divide and to spread beyond the impact crater for its descendants to infect an entire virgin world.

3

Cosmic requirements

In the beginning

Thirteen billion years ago, the youthful universe was completely dark. The searing light from the fire of its conception, the Big Bang, had long since dissipated as the universe expanded. The initial burst of pure energy condensed to form subatomic particles; protons, electrons and neutrons which, within the first few minutes, assembled themselves into the simplest elements. The Big Bang mostly produced hydrogen, some helium and minuscule amounts of lithium. The splendid Periodic Table of the Elements, containing the ninety-odd naturally-occurring elements we have discovered on Earth, had only the first three boxes filled in. Life depends on the heavier elements, not only for constructing complex polymers such as DNA or proteins but also for building the rocky planets or moons that provide a home. The early universe was not only completely sterile, it was also not even complex enough, in atomic terms, to provide the basic chemistry kit of life.

The Big Bang left its imprint on the distribution of mass, as thin wisps of matter threaded through great voids of dark emptiness. In time, these gossamer veils collapsed, under the force of gravity, into long strings of galaxies. Within these galaxies, vast clouds of gas were themselves collapsing, becoming hotter and hotter until nuclear fusion reactions flickered into action, flooding the universe with the first light it had seen since its creation. On the universe's one billionth birthday, the candles on the cake were the very first stars.

The immense heat and pressure in the cores of these first stars forced atomic nuclei together. This process, nuclear fusion, unleashes enormous quantities of energy and is the process that keeps stars burning. Even more importantly, at least for the prospects of the future development of life, fusion creates heavy elements from the primordial matter created in the Big Bang. The atoms crucial for life were formed in the nuclear furnaces in the hearts of the first stars. They are the original alchemists, converting hydrogen into the full diversity of the Periodic Table.

Hydrogen nuclei, which are single protons, are consumed first, squashed together to produce helium nuclei, with two protons. This stage of nuclear fusion proceeds until the centre begins to run out of hydrogen fuel. The star finds itself with a core of helium ash, surrounded by a layer of hydrogen that continues to burn. The great weight of the star above is no longer balanced by the outward pressure from the fusion within and the core is further compressed, creating even higher temperatures and pressures. As the core shrinks, the outer envelope of the star's atmosphere swells, producing an enormous, bloated star, many hundreds of times larger than our Sun and relatively cool on the surface. These agéd swollen stars, red giants, have a core temperature of around 100 million degrees; at this temperature, helium begins to undergo fusion. Three helium nuclei (or alpha particles as they are sometimes called) fused together creates carbon, a reaction known as the triple alpha process. The addition of another helium turns carbon into oxygen.

The triple alpha process deserves a longer explanation. The addition of the third helium nucleus is greatly enhanced by the fact that the energy of the reaction is almost exactly that of an excited state of carbon. This is known as *resonance* and means that carbon is produced by nuclear fusion much more quickly than it would be otherwise. The occurrence of this correspondence in energy levels is something of a cosmic fluke – if some of the constants of physics were even minutely different, stars would not forge carbon. In some ways, it appears as if the universe has been tailor-made, with certain factors finely tuned to allow carbon production and subsequently life. This does not at all imply the existence of any purposeful design; it is merely the consequence of observational bias. Had the universe been created with slightly different rules, that perhaps did not allow carbon synthesis, then our life would be impossible and we would not witness such a different universe. It should come as no surprise that things are exquisitely arranged to promote life, as we would never see it any other way, an idea known as the *anthropic principle*.

The helium–burning reprieve in the core of a star does not last long and fusion begins grinding to a halt again as the available helium is used up. The interior of the star is now structured rather like a hot onion, with a core of carbon and oxygen surrounded by a shell of helium and an outer layer of hydrogen. The cores of red giant stars are compressed even further and another round of fusion produces elements such as silicon, sulphur and others up to iron, which has a large nucleus with twenty-six protons. Iron is the end of the road for nuclear reactions in the star core, as its nucleus is too stable to release any more energy when it undergoes fusion. This heart of iron ultimately kills the star: the massive weight of the outer layers continues to crush the core but the compression will not be countered again by the outwards pressure of renewed fusion. The stellar collapse continues until the internal pressure reaches truly extreme levels, as the constituent parts of the nuclei, the electrons and protons, are themselves forced together. The core can compress no further and the unimaginable heat generated causes the outer layers to detonate violently outwards, ripping the star apart in the colossal explosion of a *supernova*. For a short period during its death throes, this single star outshines the rest of the galaxy. The event is so energetic that a final round of fusion reactions occurs, involving even the iron nuclei to form the heaviest elements in the universe. The outer layers of the star are hurled into space, scattering the biologically-crucial heavier elements across the cosmos. Oxygen soon combines with the hydrogen of interstellar clouds in the ratio of one to two. This is the chemical formula of water which, as far as we understand, is quite literally the liquid of life and so these very first stars gradually turned the galaxy wet. Some of the other new elements, such as copper and iodine, are used in terrestrial biology but the greatest importance of the very heavy elements is that many of their isotopes are radioactively unstable. Elements like uranium are not directly relevant to building biological molecules but they play a crucial role in the development of rocky planets. Radioactive decay helps drive volcanism and plate tectonics on Earth and may even have created some of the organic molecules that the emerging life depended on.

The first stars were extremely massive, hundreds of times fatter than our own Sun. Such heavyweights burn through their nuclear fuel at prodigious rates, quickly exhausting their resources and reaching their explosive endgame. The first stars typically lasted only a few million years – less time than our species has taken to diverge from the chimpanzees. The material blasted out by their deaths floated through interstellar space, collecting together as loose clouds, *nebulae*, of gas and dust. These clouds, loaded with the heavier elements, collapsed under gravity to form new stars. This process went through several rounds, each generation of stars processing the gas through their nuclear fires. Our galaxy gradually became richer and richer in the elements capable of interesting chemistry. The most recent stars are therefore greatly boosted in the heavier elements. For this reason, they are called *metal-rich* – to the probable distress of chemists or blacksmiths, astrophysicists call any element heavier than helium a metal.

Our own local star, the Sun, is a member of this latest breed. Our solar system formed from a nebula loaded with heavy elements, all produced by the Sun's ancient ancestors. It is true to say that we, indeed all life, are made of stardust but so too are the rocks, oceans and air of the Earth. In fact, a small proportion of the hydrogen atoms that make up the molecules of your body don't come from this galaxy; the sheer violence of supernova detonations ensures that some material is ejected right out of its home galaxy. Hydrogen from the outer envelopes of exploding stars in our next-door galaxy, Andromeda, has been picked up by the Milky Way and incorporated into nebulae here. Without wanting to overstrain the sentiment, we are not only children of the stars but also, partially, intergalactic beings.

The first generations of stars in the universe were crucial in forging the elements needed by life. The six most important are carbon, hydrogen, nitrogen, oxygen, phosphorus and sulphur, sometimes abbreviated to CHNOPS; others, like sodium and several other metals, are also crucial in small amounts. But the rudimentary building blocks of cells are organic molecules; the amino acids that make up proteins, the nucleobases of DNA or RNA and simple sugars. And it turns out that even these organic molecules could be ubiquitous in the galaxy.

Cosmic cookery

Star-forming nebulae are very large and typically contain enough mass to create millions of stars the size of our Sun. Such clouds are ninety-nine per cent gas but also contain a small amount of solids – microscopic dust particles and ice crystals. The gas is mostly hydrogen but thanks to the fusion furnaces of previous generations of stars other atoms, such as carbon, oxygen and nitrogen are also present. Over the millennia these lone atoms have bonded together to create simple compounds such as water, carbon monoxide, ammonia and hydroxyl.

The first stars to form in the nebula light it up from the inside. The Orion nebula, the middle point of the constellation's 'sword', glows a vivid red from the hydrogen it contains. The light given off by these hot young stars, in particular the UV, is absorbed by the simple molecules, together with galactic radiation, making them more reactive. Much of the interesting cosmic chemistry takes place in the denser regions around new stars, where the temperature approaches about −170 °C. The solid particles, although a tiny proportion of the total mass of the molecular clouds, play a large role in astrochemistry. Reactive chemical species can be adsorbed on to the solid surfaces, which act as catalysts, bringing different molecules together to react.

The molecules are endlessly recycled between gas- and solid-phases, evaporating back into free-floating through the nebula and re-adsorbed on to an icy dust particle. Dust particles play an important role in shielding the deep interior of the nebula from too much heat or UV radiation from the maturing stars. UV light is something of a double-edged sword – it provides the energy to allow simple molecules to bond together but is destructive of more complex molecules.

Over time, more and more complex organic molecules build up within the dust clouds, although always a thousand-times less abundant than the hydrogen. Astronomers have detected around 130 different molecules, the most complex of which are composed of as many as thirteen atoms bonded together. Many of these, such as ammonia, formaldehyde and hydrogen cyanide, are thought to have been very important in the prebiotic chemistry in Earth's seas. The most common organic molecules in these interstellar clouds are polycyclic aromatic hydrocarbons (PAH), which have a backbone of carbon atoms arranged in a variety of slightly different configurations. One example is naphthalene, the chemical that gives mothballs their distinctive smell. They are the sort of compounds that escape in soot particles from car exhausts and are not themselves important to terrestrial life. But it is very easy chemically to convert PAH into the porphyrins and quinones that form the metal-holding cores of many biological molecules, for example metabolic enzymes, the active centres in chlorophyll and haemoglobin and components of the electron transport chain. The formation of these precursors is central to the development of life on Earth.

Although many different simple organic compounds have been found in these great clouds of gas and dust, nothing as complex as even the simplest amino acid has been unambiguously detected. Perhaps these complex compounds are just harder to find because they are relatively rare and must be deep within the cloud to be protected against destruction by UV radiation. Lab experiments that attempt to recreate the cold gas, dust particles and UV levels of nebulae have found that glycine, alanine and serine – the three most abundant amino acids in earthly proteins – can indeed be produced. However, the best evidence for the cosmic origin of more complex organics comes from a certain class of space rocks.

Carbonaceous chondrites are a type of rock from the asteroid belt. They contain a few per cent of carbon compounds and are believed to be left over from the primordial material the planets were made from. One example, the Murchison meteorite, fell to Earth in Australia in 1969 and has been extensively analysed. The diversity of organic molecules it contains is truly staggering: various sugars, the five nucleic acid bases used by terrestrial DNA and RNA and chains of fatty acids which, when mixed with water under the right conditions, spontaneously form into hollow vesicles roughly the size of bacteria. But perhaps the most surprising find is the more than seventy amino acids. Terrestrial proteins use twenty different amino acids, of which only six were found in the Murchison meteorite, so the vast majority are utterly alien to life on Earth. This last finding is not the only convincing proof that the organic substances present aren't simply contamination from terrestrial life since the meteorite landed. The sugars and the amino acids in the rock are of both enantiomers which, as I described in Chapter 1, is a sign of abiotic chemical reactions. The enzymes of life on Earth produce only one enantiomer of an organic molecule. However, there does seem to be a slight, but none the less intriguing, bias towards the left-handed enantiomer in the Murchison amino acids. There is no clear idea about how this bias arose but it is obviously of great interest in working out why terrestrial life favours one enantiomer over the other. One theory is that the complex amino acids formed within the interstellar clouds, before the birth of the solar system and were affected by the polarisation of the UV light from nearby stars. If true, it is possible the enantiomer bias displayed by cells on Earth is an imprint of this ancient starlight.

The galaxy seems to be a very encouraging place for astrobiology; the building blocks of terrestrial biochemistry are apparently widespread through the cosmos. How all these raw ingredients arrived on Earth is a topic I shall cover in the next chapter. For now, we'll look at what sort of planet is needed to nurture this prebiotic soup into living matter.

A suitable world

As I discussed in the first chapter, liquid water is thought to be absolutely necessary for life. Practically every moist niche on Earth has been colonised by prokaryotes and although water alone is not enough for life to evolve, it's certainly thought to be a crucial prerequisite. For a world to evolve life, it must, first and foremost, have expanses of liquid water. This was at first thought to require a large rocky planet with a hot core, thick atmosphere and surface oceans – a terrestrial planet. I shall discuss the possibility of life on worlds other than warm rocky planets later but for the moment, I shall look in detail at what Earth-like life probably needs.

The factors that control whether ground water is liquid are the surface temperature and pressure of the planet. The conditions under which water flows can be plotted on a graph with these as the axes, as in Figure 5. If we look half way up, at one atmospheric pressure – sea level pressure on Earth – then we can trace our eyes horizontally across for increasing temperature. We see the transition from solid to liquid, melting, at the familiar 0 °C and boiling at 100 °C. Increase the pressure, such as at the bottom of the ocean

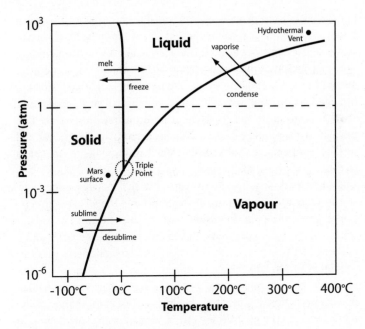

Figure 5 Phase diagram of water showing the temperature and pressure regimes where it is liquid. Below the triple point fluid water is not stable and sublimes directly into the vapour phase.

and water is still liquid at over 300 °C. On the other hand, if you decrease the pressure below the triple point (the temperature and pressure at which all three phases – gas, liquid and solid – of water may co-exist) then water cannot exist in a liquid state at all; it sublimes straight from solid to gas. This is the present situation on Mars; we'll explore the implications of this in Chapter 5.

If, for the moment, we assume that an alien world has an atmosphere as thick as our own we can see what range of temperatures will support liquid water. The surface temperature is largely determined by how close to its star the world orbits. Too close to the central heat, and water boils off; too far away and the water is permanently locked up as ice, unavailable to perform biological functions. The range of distances that allows liquid water is known as the stellar habitable zone (HZ). This is like a thin ring circling the star, within which the planet must orbit. It is sometimes referred to as *Goldilock's Principle*: life needs a planet that is not too hot, not too cold but just right. However, this is a very simple look at what determines the presence of liquid water on a world. The planet's atmosphere plays an enormous role in regulating the surface temperature and pressure. We'll now turn our attention to this.

Earth's thick atmosphere insulates the surface very effectively. Carbon dioxide, methane and water vapour in the air act as 'greenhouse' gases, allowing the visible part of sunlight through to heat the Earth's surface, so that it begins to give off infra-red, or thermal, radiation. But such gases absorb these lower wavelengths of light, trapping the warmth below; an effect identical in principle to the panes of glass trapping heat within a garden greenhouse. The greenhouse effect is a perfectly healthy part of a planetary system: if it weren't for the greenhouse effect, life on Earth's surface would be impossible, as the average temperature would be more than thirty degrees lower and the surface a frozen wasteland; the concern is that humans are now pumping so much carbon dioxide into the atmosphere that it might begin tipping the climate away from its current, stable, point.

The carbonate-silicate cycle

As I briefly touched on in the introduction, the carbon dioxide in the atmosphere is not static but is one stage in a great loop of carbon moving through the rocks, water and air of the planet, properly called the *carbonate-silicate cycle*.

Dissolved in water, carbon dioxide forms a mild acid, *carbonic acid*, which weathers silicate rocks to form the mineral calcium carbonate, or chalk. This rock becomes deposited on the ocean floor, effectively locking away the carbon. Without another process, by which the carbon is returned to the air, all carbon dioxide would be scrubbed from the atmosphere in a few hundred million years. This would be catastrophic for the climate, as an important greenhouse gas steadily disappears and the world freezes. The return leg of the cycle is provided by plate tectonics.

On Earth, new crust produced at the ocean ridges migrates slowly outwards, accumulating thick layers of carbon-containing sediment. The crust is pushed inexorably towards *subduction zones*, where it is forced downwards, to melt in the hot interior mantle. The carbon dioxide is released again and vented into the atmosphere through volcanoes. Most crust completes the journey in about 150 million years, steadily recycling carbon dioxide back into the air.

Most importantly, this particular elemental cycle is self-regulating. The rate of weathering of silicate rocks to form chalk is temperature-dependent. The warmer the conditions, the more quickly carbon dioxide is removed from the atmosphere, reducing the greenhouse effect. In a colder climate, the reaction slows and carbon dioxide accumulates, raising the temperature. This negative feedback automatically keeps carbon dioxide levels in check; as though the entire planet has a thermostat that very carefully regulates the global climate. The oceans are not only vital in providing an environment for life but the water also performs a crucial role in controlling the temperature, eroding silicate rocks and lubricating the motions of the tectonic plates over one other.

The stellar habitable zone is determined by a complicated interplay between the brightness of the star and the insulating properties of the planet's atmosphere, regulated by the carbonate-silicate cycle. Seas of liquid water and carbon dioxide in the air are critical to the functioning of this climate control system; these two factors really define the inner and outer boundaries of the HZ. The position and width of the HZ calculated for different kinds of stars will be shown in Figure 15 when we return to this important concept in Chapter 7.

Runaway greenhouses and glaciers

The process that acts to lower the surface temperature can only do so much. If the planet is too warm, large amounts of water vapour, which is itself a greenhouse gas, evaporate into the air. But unlike carbon dioxide, water vapour is under no negative feedback control and the world gets hotter and hotter. Eventually the atmosphere becomes laden with steam, as the oceans boil dry. Without the lubricating effects of the water, the conveyer belt of plate tectonics creaks to a halt and the carbonate rocks exposed on the surface decompose in the heat to release their carbon dioxide. UV light from the star splits apart (photodissociates), the water vapour high in the atmosphere and the lighter hydrogen gas escapes the planet's gravity completely. The planet's inventory of water is lost forever and thus its potential for developing life. The distance from a star where this catastrophic, runaway greenhouse process is triggered defines the inside limit of the HZ.

On the other hand, if a planet is slightly too far away from its star, even a thick atmosphere cannot provide sufficient warming. The oceans begin to freeze, creating wide areas of pure white on the planet's surface that very effectively reflect more of the sun's warmth. The global temperature nosedives; eventually even carbon dioxide begins to freeze out of the atmosphere. With this important greenhouse gas deposited as a frost on the ground, all hope of climatic recovery is lost. Such a fate, runaway glaciation, sets the outside limit of the HZ.

For a planet to remain pleasant long enough for life to emerge it must skirt gingerly between the twin dangers of runaway greenhouse and glaciation; Earth has obviously managed this for the last four billion years. Computer simulations of Earth's climate have been built to calculate the extent of the Sun's HZ. The average distance of the Earth's orbit from the Sun is known as the astronomical unit (AU). Depending on what assumptions are made about the various factors, estimates for the current HZ range from a generous 0.7–1.2 AU, to a very narrow 0.958–1.004 AU. Either way, it is clear that the habitable zone around a star is very thin indeed, a knife-edge on which Earth is precariously teetering.

The type of orbit the planet follows is also important. Earth's route around the Sun is almost circular but could life be sustained on a planet with a much more elliptical path, one that swoops close in to its star before swinging further away? Leaving aside that this may bring the planet too close to the gravitational effects of others, possibly flinging it out of the HZ altogether, such an 'eccentric' orbit probably poses harsh climatic problems. The meandering world would edge towards a runaway greenhouse effect on one part of its path and runaway glaciation during the other. Even if the entire orbit is still within the bounds of the HZ, the wild swings in heat input might make for an erratic climate, preventing the stable conditions necessary for life. However, large oceans might act as an effective temperature buffer, slowly warming up and storing the excess heat of summer for release during the outer region of orbit.

The development of a solar system with a rocky planet within the HZ of the star is only part of the problem. The bounds of the HZ are not static but move as the star matures and grows steadily hotter. Over time, the ethereal ring of safety around the star sweeps outwards. For example, over the Earth's history the Sun has brightened by about twenty-five per cent but the average global temperature has remained remarkably constant, thanks to negative feedback loops like the carbonate-silicate cycle. However, this negative feedback can only compensate for so much. At some point in the Earth's future the ageing Sun will brighten to such a degree that the inner edge of the HZ will sweep right past our planet. The atmospheric thermostat will break down and the inexorable process of runaway greenhouse be set in motion. This fate won't befall the Earth for another billion years or more but the rate of solar brightening depends on the size of the star. The more massive a star is, the more quickly it burns and therefore the less time a planet has before it slides out of the habitable zone. This is the concept of the continuously habitable zone – not only must a planet be in the right orbit but its parental star must also be of a suitable, long-lived kind. Terrestrial planets orbiting stars only twenty per cent or so more massive than the Sun are unlikely to have enough available time to evolve life. Stars smaller than the Sun smoulder much more slowly and can provide stretches of time ample for life to develop. However, such cooler stars have many of their own problems, which I shall discuss in Chapter 7 and it is not yet clear whether they are feasible astrobiological targets.

As well as a stable orbit within the continuously habitable zone of its star, size really does matter when it comes to the habitability of a planet.

Neither dwarf nor giant

In some ways, the Earth is a bit of an oddity in the Sun's family of planets. It is the largest rocky body in the solar system; it is also the only one with a detectable, planet-spanning, ecosystem. Since (so far) we have detailed information about only this one solar system, we cannot tell whether this is merely a coincidence or if the Earth really is abnormally large; large enough to have permitted the development of life. Astrobiologists have some very good reasons to think that life might require a roughly Earth-sized planet.

A planet that is much smaller than Earth would have too feeble a gravitational grip to cling on to a thick atmosphere for the aeons of time necessary to evolve life. A thin atmosphere would provide only limited greenhouse heating and the lower air pressure would limit the temperature range of flowing water. Thus, a small world would face difficulties sustaining liquid conditions long enough for life to evolve. Small planets also lose their internal heat much more quickly. The Earth's core is still very hot from its formation and the mantle above is fluid enough to allow the convection currents that drive plate tectonics. This geothermal energy ensures that the Earth continues to be very volcanically active. Plate tectonics and volcanism are critical for the carbonate-silicate cycle that keeps the global temperature stable.

There is another important function of the Earth's hot interior: the churning iron within its core induces a powerful magnetic field. This forcefield reaches high above the top of the atmosphere and safely deflects the solar wind, a fast flow of particles streaming off the surface of the Sun. On planets without this protective cocoon, the solar wind slowly but steadily blows away the atmosphere. Mars is believed to have lost a large volume of its original gases and its water into space partly because its magnetic field failed almost instantly after the formation of the planet. Thus, for several reasons, planets with a thinner atmosphere and cooler interior would be much more vulnerable to the threat of a runaway glaciation.

Very small planets would not have been hot for long enough to create a differentiated interior. The inside of the Earth is layered; the heaviest elements, such as iron and nickel have sunk to the core, which is covered by a mantle of molten silicate rocks underneath a thin, solid, crust. Some of the energy that drives tectonic shifts and volcanism comes from the heat released by the decay of radioactive isotopes of uranium, thorium or potassium, the heavy, unstable nuclei produced in previous supernovae. For a planet to be geothermally active for long periods, these radioactive elements must be concentrated in the core, otherwise their heat would be rapidly dissipated.

A dwarf Earth would suffer several problems in developing life. Its lower gravity and thicker, more rigid, crust would allow higher mountains but the loss of atmosphere and water would mean cold surface conditions and shallow seas. Runaway glaciation is a likely destiny for such a world. A giant Earth would encounter two main problems. First, the hotter interior would mean high volcanic activity, with a continual spew of greenhouse gases out into the atmosphere, causing higher surface temperatures. (The greater gravitational pull of a large planet would also retain a much thicker atmosphere, again contributing to higher temperatures.) Second, the planet might run into serious trouble with its own geography. A hotter interior means a thinner, more pliant crust which, combined with the stronger gravity, would mean that the surface would be relatively smooth, with little relief. The higher gravity (and probable strong magnetic field) would mean little of the atmosphere or water would have been lost. These conditions would favour a very deep, planet-wide ocean, with little, if any, dry land. The lack of exposed rock would mean a very low rate of silicate erosion. The carbonate-silicate cycle would be broken and with no effective mechanism to remove excess carbon dioxide from the atmosphere, its levels would rise inexorably. Thus a large terrestrial world would seem to provide the perfect conditions for a runaway greenhouse effect, ultimately boiling off the oceans and sterilising the face of the planet of any life that may have emerged.

So, the stellar HZ and the details of the planet's atmosphere, which is itself largely dictated by the mass of the world, are closely intertwined. Our two neighbouring planets, Venus and Mars, perfectly illustrate this point. Venus has a heavy atmospheric blanket of carbon dioxide and a hellishly hot and bone-dry surface. It is a clear case of a world with a runaway greenhouse. Mars is bitterly cold, with a paper-thin atmosphere and an easy diagnosis of runaway glaciation. Intriguingly, if the positions of the two planets had been swapped at the birth of the solar system, with the smaller Mars closer to the Sun and Venus further away than Earth, their development may have turned out very differently. Mars, despite its thinning atmosphere and reduced volcanism, may have stayed warm enough to retain seas on its surface and Venus may never have tipped over the threshold that sent its greenhouse spiralling out of control. Instead of just Earth, the solar system might have had a family of three habitable planets.

The Moon

The Earth has one other quirk that may prove to have been important for the survival of life, particularly multi-cellular animal life: the largest moon of all of the inner planets. Neither Mercury nor Venus have moons and Mars' two companions, Phobos and Deimos (the god of war's attendants, the spirits of Fear and Panic), are not thought to be true moons but captured asteroids. Phobos is in an unstable, decaying, orbit, spiralling in towards the Red Planet as it loses momentum. Calculations show that in about forty million years' time, Phobos will begin brushing through the upper reaches of the Martian atmosphere and from there will, literally, fall out of the sky. A similar impact on Earth would trigger mass extinction.

There is also an uncanny coincidence about the Moon – it appears, in the sky, to be almost exactly the same size as the Sun. (This means that the Earth is the only place in the entire solar system where one can watch a total solar eclipse, with the characteristic thin ring of fire around the edge of the Moon, where the Sun's upper atmosphere is just visible.) If the Moon is crucial for complex life, its importance will be its tidal effect on Earth. For a moon is to draw significant tides, compared to the effect of the Sun's gravity, it can be shown that it must have an apparent size similar to, or larger than, the Sun's.

The main claim to importance of Earth's large Moon is that its gravity stabilises the spin of the planet. The cycles of day and night are caused by the Earth rotating around its (pole-to-pole) axis. This axis is not perpendicular to Earth's orbit but tilts by 24°. This angle of inclination is affected by the gravitational tug-of-war between the Sun and the giant gas planets, Jupiter and Saturn. The effect of the nearby Moon is to overpower these other influences and keep the Earth's tilt much more stable. There is evidence that, in the past, Mars, which currently has a tilt very similar to Earth's, slanted by 60°. Without the Moon, Earth's lean would sway chaotically; calculations have shown that it could approach an incline of 85°. We know little about what the effects of such a slant would be. Computer simulations of the climate without the Moon's influence have been run, gradually toppling the entire planet head-over-heels until the poles are almost horizontal. Over a year, the poles would receive more sunlight than the equator and so the climatic conditions in these two regions might, effectively, be swapped. It is not clear whether a swaying planet would be significantly more prone to runaway climatic processes. Complex land life, like Earth's, would probably be impossible but we do not know whether it would matter to life developing deep underground, for example.

Guardian Jupiter

It has been claimed that a large gas giant planet in the outer solar system is a prerequisite for life to evolve on a rocky planet within the stellar HZ. Impacts with comets or asteroids are known to be serious hazards to a planet's biosphere – the one that hit Earth 65 million years ago is almost certainly the culprit in the extinction of severty-five per cent of its species, including the dinosaurs. The premise is that the gravitational influence of a gas giant clears away stray comets, either taking the hit themselves (like the renowned collision of Comet Shoemaker-Levy 9 into the Jovian clouds in 1994) or deflecting them out of the solar system. Jupiter might be a celestial goalkeeper, protecting the inner planets from potentially devastating impacts. It has been estimated that, without the guardian effect of our bigger brother, a thousand times more comets would have fallen Earthward. However, this presumed necessity is not as convincing as the others we have already explored. Cometary impacts are not always entirely bad. If it hadn't been for the abrupt end to the reign of the dinosaurs, our shrew-sized mammalian ancestors would never have had the evolutionary opportunity to spread and eventually produce our conscious and technologically-able species. Moreover, as we'll see in more detail in the next chapter, life on Earth would be utterly impossible if it hadn't been for a bombardment of brutal collisions during our planet's childhood that delivered not only the oceans but also all the organic molecules manufactured in nebulae.

It has also been claimed that Jupiter disrupted the growth of the inner solar system and, had it not existed, both the asteroid belt and Mars would have formed planets larger than Earth. A more massive Mars could have retained a thicker atmosphere and warmer surface and so may have remained habitable. Others think that the gravitational disruption of a large gas planet is crucial to speeding up the formation of the inner planets from colliding rocky embryos, as I shall describe in the next chapter.

So, a planet must be of the right size, orbit a suitable star at the correct distance and have several other appropriate features of the solar system. The prospects for a galaxy teeming with life get even worse when we consider that, within a galaxy, only certain regions may be suitable.

A galactic habitable zone

Our galaxy, the Milky Way, is an immense rotating spiral of around 400 billion suns, about 100,000 light years across. Many of these stars can be discounted as serious contenders for life, as they are too fleeting, and many more may have no terrestrial planets within their habitable zone.

As I have described, metallic elements are important for life. Cells depend on the complex chemistries of a wide range of heavy elements. The rocky terrestrial planets are essentially great lumps of iron, silicon, oxygen and the radioactively unstable isotopes of uranium, thorium and potassium needed as a heat source deep within the planet's interior to drive plate tectonics and volcanism. Due to the development of the galaxy, it turns out that these heavier elements are not scattered evenly but are concentrated in the centre. There is a 'metallicity gradient', running from the galactic core to the outer rim. This means that stars towards the outside are thought to be unable to form a family of planets and so cannot provide even the basics for terrestrial life. Our Sun, perhaps unsurprisingly, lies about half-way between the galactic centre and its edge. The decreasing abundance of heavy elements may mark an outer boundary for the distribution of life, beyond which planets cannot be assembled. The inner edge of such a galactic HZ is less well defined but it is thought to be no coincidence that the Sun lies no closer to the centre of our galaxy. Certain factors may prevent life getting a foothold within the inner regions as great threats to the continuation of life lurk throughout the galaxy.

Killers from outer space

Supernovae explosions are crucial for the creation and scattering of heavy elements but their sheer violence also makes them a considerable hazard to life-bearing planets, even across the gulfs of space. The greatest threat is posed by supernovae around thirty light years away. (This is a mid-point between the devastating effects of a very close detonation and the decreasing probability that a star is that near to Earth when it explodes.) A star exploding thirty light years away would hang blindingly bright in our skies but it would not be the extra light that spelt doom to ecosystems on the surface. The detonating core of the supernova would greatly accelerate material in the expanding shell, creating a pulse of highly-energetic particles. Radiation like this is spread throughout the galaxy; cells on Earth have evolved mechanisms to repair damage caused by low levels but a nearby supernova explosion would strike the Earth like a point-blank shotgun blast. As well as the direct problems of high mutation rates in exposed organisms, this surge of radiation would destroy the ozone layer for centuries and allow UV light from the Sun to stream down on to the surface, damaging the delicate molecules of life. The machinery of photosynthesis is particularly sensitive to UV light; a nearby supernova could shut down photosynthesis both on land and in marine environments and the ecosystems dependent on it would crumble. The long-term effects of such a close explosion are less clear. The radiation would increase cloud formation, reflecting more sunlight and possibly triggering runaway glaciation. Alternatively, the sudden drop in photosynthesis and reduced removal of carbon dioxide from the atmosphere might tip the balance the other way, increasing global temperatures.

Only large, and thus relatively rare stars, bigger than about eight Sun masses, result in supernovae. Although a supernova occurs somewhere within our galaxy every few decades, the chance of Earth being within thirty light years when one goes off is very small. However, it is likely that at least one close detonation has occurred during the history of complex animal life on Earth. Since the evolution of hard-shelled animals 550 million years ago, there have been several major mass extinctions. The event, 65 million years ago, that wiped out seventy-five per cent of Earth species, including the dinosaurs, has confidently been attributed to a massive impact. But out of the 'Big Five' mass extinctions this was a relatively subdued catastrophe and no convincing explanations have been found for many of the others. It is possible that at least one of the great disturbances to life on Earth may have been triggered by a supernova. However, if this were the cause of a previous mass extinction we would be unlikely ever to find the smoking gun. The relative motion of stars means that, after swinging towards Earth and exploding, the supernova remnant could now be on the far side of the galaxy. A layer of a heavy isotope of iron in the deep ocean crust is evidence of a nearby supernova explosion in the last few million years – when the first human-like ancestors were evolving. The explosion occurred at about 100 light years range and we can still see the débris drifting away in our neighbourhood.

The Sun's orbital path through the galaxy is relatively unpopulated with stars. The threat from a nearby supernova would be greatly increased if the solar system were instead located in a region with a much greater star density, such as towards the galactic core. And the tighter packing of stars in the inner districts of the galaxy is dangerous for a second reason: close encounters with other stars would be much more likely, with the errant gravitational tug playing havoc with the solar system. The paths of the planets might be nudged enough to trigger wild climate swings but it is certain that the swarm of comets lurking in the outer reaches of the solar system would be scattered, sending some plummeting towards the inner planets. Earth would be subject to a firestorm of impacts, with grave consequences for the survival of life.

Although the Sun is currently passing through a relatively uncrowded area of the galaxy, we are by no means completely in the clear. The Milky Way has a rotating, spiral, structure, much like the white foam swirling around on a fresh cup of coffee. If you watch your cappuccino carefully, you will notice that the inner regions turn much faster than those further out; likewise, stars within our galaxy constantly overtake others on their outside. The four spiral arms do not turn as rigid formations, they're just regions where lots of stars are born and light up the arm before they migrate away. The Sun's orbit around the galactic core regularly takes it through one of these spiral arms, about every 100 million years or so. Each traverse takes about ten million years; throughout which time the solar system runs a gauntlet of dangers.

The spiral arms are dense concentrations of stars and so present an increased supernova risk. A second hazard is the great dust clouds that give birth to new stars. These nebulae are concentrated in the spiral arms and can be hundreds of light years across. The solar wind would generally be strong enough to blow aside the cloud without any ill effect but an especially dense section of cloud would overpower the Sun's protective bubble and flood into the solar system. The dust would gravitate towards the planets and settle in the upper atmosphere, blocking out the sunlight for around 200,000 years while the solar system passes through the cloud. At the least, this could shut down photosynthesis, at the worst, it could kick-start global glaciation.

So, the spiral arms pose three main dangers – supernovae, close shaves with other stars and interstellar dust clouds. It is difficult to calculate exactly the Sun's path around the galaxy but some researchers have attempted it, finding that several mass extinctions do indeed correspond with previous crossings of spiral arms. The Sun's orbit around the galactic core is surprisingly circular, ensuring the solar system does not periodically plunge deeper towards the centre, into the dangerous regions with higher densities of stars. The orbit is also uncannily close to the *co-rotation cycle* (where the Sun's orbital rate about the core is almost identical to the rotation of the spiral arm pattern) resulting in the minimal number of crossings. This is the final feature of what many astronomers call the galactic habitable zone. Just as life can only persist on a planet at a specific distance from its star, it seems like the star itself must orbit within a certain ring around the galactic core. The outer edge of this ring is set by the minimum metallicity required to form rocky planets, the inner boundary by the various hazards posed by the central realm of the galaxy, such as the greater danger of close shaves with other stars and nearby supernovae. Even a solar system further out in the disc is not safe, as regular crossings of the spiral arms carry similar risks. The safest place for life is probably within a thin halo around the co-rotation cycle, not far off the orbit of the

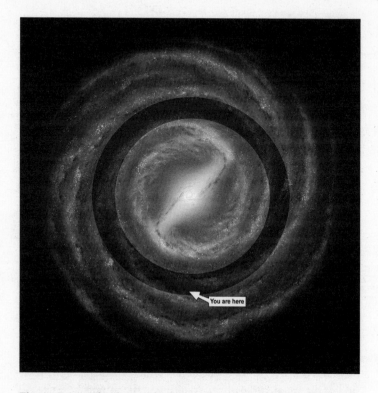

Figure 6 A simulated view of our own galaxy, the Milky Way. The approximate extent of the galactic habitable zone and the Sun's current position are also shown.

Sun, as shown in Figure 6. Some researchers calculate this haven contains a mere ten per cent of the stars in the galaxy, of which only a fraction are Sun-like.

So, it seems life may be particularly fussy about where it arises. It demands a fair-sized planet, set in the Goldilock's zone around a well-behaved star and in a convenient region of the galaxy. Not only that, whole galaxies might be of the wrong kind. Metallicity is important for constructing life-capable planets; only a portion of our spiral galaxy is suitably metallic. The metallicity of a galaxy is associated with its luminosity and it has been estimated that eighty per cent of stars in the visible universe reside in galaxies less luminous than our own, suggesting that most stars are relatively metal-poor and may be unable to form planetary families. Within our local group, including the Milky Way, Andromeda and twenty-four other galaxies, most are either blob-like or irregular and metal-poor. Stars in these types of galaxies also have much more erratic orbits and regularly plunge in towards the dense galactic core, with all the dangers that entails. Our Sun might be something of a rarity, not only within the Milky Way but also compared to the majority of other galaxies.

These conclusions are very dependent on the range of conditions within which life might be able to survive; conditions researchers are striving hard to understand. Certain issues I have discussed, such as the greater threat of a close-quarters supernova nearer the galactic centre, would not prevent life from arising but may present a hazard to its persistence, particularly the persistence of more complex life. To the best of our knowledge, although obviously currently based on only a single datum point, the requirements I have described in this chapter are thought to give life its most favourable opportunity for developing and surviving. In later chapters I shall relax these constraints and see what other galactic real estate might be suitable.

4

Earth's story

Accretion: the birth of a star and its planets

The heavy elements necessary for life, for building suitable rocky planets and complex polymers, are forged in stellar furnaces and redistributed into interstellar nebulae. But how are these great clouds of gas and dust moulded into stars and planets? The key is, once again, found in exploding stars. The shock wave from a nearby supernova explosion compresses a region of the nebula enough for it to begin to collapse under its own gravity. The contracting gas, which had been eddying only gently, rotates faster and faster to conserve momentum, just as an ice-skater spins more rapidly when she draws in her arms. The violently-swirling cloud breaks up into smaller vortices, which will form into individual solar systems. Depending on the size of the fragments of rotating gas, they will either form an isolated solar system with a single central star or clusters of multiple stars whirling around each other. The original explosion, in one go, creates a collection of stars that sparks into light several million years after the supernova event. Our Sun, although created in the same event as many others, was born a singleton and does not orbit in a cluster of stars, something we'll return to in Chapter 7.

As the individual vortex of turbulent gas that will become our solar system collapses, it flattens into a rotating disc. The central region, which will give birth to the Sun, is surrounded by a skirt of material, the *accretion disc*. The total mass of this disc is only about a hundreth that of the Sun but more than enough to make a family of planets, asteroids and comets. This accretion disc stretches ten times beyond the current orbit of the dwarf planet Pluto. Ninety-nine per cent of the disc is gas; the rest tiny particles of floating dust at a density of only a few per cubic metre.

The swirling dust particles stick together, growing into larger and larger lumps. Once they become large enough to exert an appreciable gravitational tug, their growth rapidly accelerates, the rich getting richer as they pull in more and more material. Within about a million years, the dusty disc accretes into several hundred planetary embryos, somewhere between the Moon and Mars in size. The gravitational interactions amongst the swarm of orbiting bodies sometimes swing an embryo into the Sun or catapult it out of the solar system altogether. The embryos occasionally collide, in crunching, shattering impacts, building the rocky inner planets we know today. The heat of these impacts and that from the decay of radioactive elements in the rock melts the interior of the growing planets and the heavier elements, such as iron, settle into the core. These terrestrial planets are, however, almost completely devoid of carbon or water, two substances thought to be essential for life. This is because they form in the central region of the solar nebula, inside what is known as the ice-formation boundary or 'snow line'. This inner zone, extending out about five astronomical units, is too warm for volatile compounds such as hydrogen, carbon monoxide and water to become incorporated into the growing dust grains. Such volatile substances will be delivered to terrestrial planets later in the development of the solar system, a critical process we'll look at much more closely soon. The growth of embryos into full-blown terrestrial planets takes about a hundred million years.

Meanwhile, back in the outer regions of the solar nebula, the gas giant planets are forming. There, beyond the snow line, temperatures in the accretion disc are below $-120\ °C$ and the volatile gases condense into crystals. The extra material out in these suburbs results in the formation of much larger planetary embryos. Jupiter, for example, has a core of rocky silicates and ice about thirty times larger than Earth's. The gravitational attraction of such enormous bodies is strong enough to draw in large amounts of the nebula gas, mostly hydrogen and helium. This in turn makes the planet more massive, so that it pulls in extra gas, a snowball effect that quickly produces truly gigantic gas worlds. Jupiter, king of the planets, will grow to be 318 times heavier than Earth. Even further out in the developing solar system, there is little material available to build planets and only small icy bodies, like Pluto or comets, form.

While the accretion disc evolves from a swarm of embryos to a mature family of inner terrestrial planets and outer gas giants, the Sun undergoes its own development. As the central zone of the nebula contracts, both its density and temperature increases. After about half a million years, the proto-Sun becomes hot enough for the first stages of nuclear fusion to light up the inside of the cloud. It is quite inflated, some five times larger than its current size, but shrinks as more of the surrounding gas piles down on top of it. Eventually, the density and temperature in the core are pushed over a critical threshold, around ten million degrees Celsius and the main nuclear reaction, hydrogen fusion, ignites. Some forty million years after its initial collapse, the Sun reaches the end of its infancy. Today, the Sun is about half-way through the stable phase of its life, steadily brightening as it approaches the inevitable end point of red gianthood.

Moon-forming impact

The Earth formed about four and a half billion years ago, in an orbit perilously close to that of another, smaller, young planet, about the size of Mars. Almost immediately, these two neighbouring worlds cracked into each other. Half Earth's mantle was explosively ejected into space and a tsunami of molten rock raced around the globe. The smaller planet came off much worse. It was completely destroyed, its metallic core merged with Earth's and its outer rock splashed into space. Most of this material rained back down on to the Earth's surface but some remained in orbit. The ejected material may initially have looked rather like the rings of Saturn, albeit composed of shards of rock rather than glittering gems of ice, circling a glowing red-hot Earth. The pieces of the two shattered planets gradually collected into a set of small moonlets, then coalesced into the single lunar body we are familiar with today. The results of a computer simulation of the first day after this impact are shown in Figure 7.

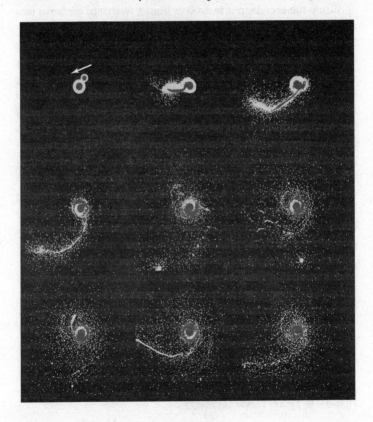

Figure 7 Computer simulation of the first 24 hours after the 'big splat' collision that created the Moon. Iron rich cores shown as dark, silicate mantle as light.

A few thousand years after this big splat, the Earth's global ocean of molten magma had cooled enough first to develop a rind and then to solidify into crust. Over time, probably by around 4.4 billion years ago, the world became cool enough for water vapour to condense out of the air. Great oceans formed, broken only by tiny continents – an estimated twenty times less land surface than now. The high internal heat fed huge volcanoes, whose peaks emerged above the waves like the Hawaiian islands, venting out large volumes of gas and cloaking the planet in a thick atmosphere.

In the early days, the Moon was much nearer to the Earth, looming in the sky twice as large as it does today, raising enormous tides over the tiny continents. Over time, this has sapped the Moon's orbital energy and it is steadily drifting backwards. Tidal friction has also drained energy from the rotation of the Earth and it spins much more slowly now – just after the formation of the Moon, a day lasted only five hours.

It is possible life developed even this early, although we will never know. No rock from this earliest period of Earth's history has survived, destroyed long ago by the planet's active surface. The Earth's crust is segmented into rigid slabs; when these tectonic plates collide, one is pushed beneath the other and melted in the heat of the interior. The volcanic activity triggered above this subduction zone produces granitic rocks, much lighter than the original crust. These new, more buoyant, blocks were not subducted, instead merging to produce the first *continental crust*. Water collected over the heavier, lower-lying rocks, of the *oceanic crust*. The oldest rocks on Earth come from these early continents, where rivers ran over the stark bare rock, leaching out the minerals and washing into the seas to turn them salty.

Heavy bombardment

So, the early oceans sloshed back and forth under the powerful tidal action of the nearby Moon. But if none of the terrestrial planets had any volatile substances, because they formed within the snow line, where did the oceans' water come from? The answer lies in the swarm of bits and pieces left over from the creation of the planets. The outer asteroid belt and comets formed beyond the snow line and so are rich in the volatile chemicals required by life. During the earliest stage of the solar system, a barrage of these bodies rained on the inner planets. It's not quite clear whether this bombardment was a tailing-off after the violent formation of the planets or whether there was a sudden volley of incoming projectiles after a period of relative peace. If the latter, such a *late heavy bombardment* may have been caused by projectiles slung Earth's way either by the late formation of the gas planets Neptune and Uranus, a slight shift in Jupiter's orbit or by the passing of a nearby star through the stellar nursery that created the Sun. This sharp blitz is believed to have run from 4.2 billion years ago with, increasingly infrequent, large impacts continuing until perhaps 3.8 billion years ago. Barely any of the Earth's original surface from this period remains, for it has either been eroded or dragged into the Earth's interior by plate tectonics but the pristine conditions of the Moon preserve a long crater record. The number and size of the Moon's craters from this era, some the size of continents, are a record of the fierce blizzard of rock and ice that must have rained down on all the inner planets. This influx of comets and meteorites delivered enormous amounts of vital volatile subtances to the Earth's surface: water to fill the oceans, carbon dioxide to warm the atmosphere and the simple organic molecules formed in the nebulae rained down like manna from heaven. Hundreds of tonnes of organic carbon is still delivered to Earth every year; the rate would have been hundreds of times higher in the youthful solar system.

The bombardment was absolutely crucial for the development of life on Earth. Without these deliveries of volatile substances from beyond the snow line, no organic molecules or oceans would have ever existed on Earth's barren surface. Life may even have gained a foothold relatively quickly after the formation of the Earth and Moon, only to be snuffed out by the barrage of a late heavy bombardment. Although the vast majority of impacts were small, every now and then the Earth stumbled into a giant lump of rock. Objects hundreds of kilometres across occasionally fell from the sky and dumped a colossal amount of energy on to the Earth's surface. Such a large asteroid would punch through the solid crust and into the fluid mantle beneath, throwing up a great splash of molten rock that would rain back to Earth, covering the surface with hundreds of metres of solidifying débris. A thick atmosphere of rock vapour forms, trapping the heat, so the ground is a searing inferno. Under these conditions, the oceans boil bone-dry in just a few months. The air remains heavy with steam for centuries afterwards, exerting a powerful greenhouse effect and baking the ground to temperatures of around 2,000 °C, destroying any trace of life or accumulated organic molecules. Eventually, this excess heat radiates away and the atmosphere steadily cools, until droplets of liquid water form. It begins to rain on Earth again, a continual deluge that lasts thousands of years, as the oceans refill.

The emergence of life

Such devastation may have befallen the Earth five or six times during the late heavy bombardment. After each mega-impact the oceans refilled, organic substances delivered from outer space accumulated and, perhaps, replicating molecules emerged and cells took a tentative foothold. But with the next cataclysmic spasm life was mercilessly snuffed out once more. This knocking-back, or *impact frustration*, limits the earliest time life could possibly have arisen – and persisted – on Earth. However, as the bombardment eased off, the impacts became less frequent and less severe. The easing of the blitz brought relative stability to the planet and life finally had a fighting chance. It's startling just how quickly life seems to have sprung up. The last sterilising impact is believed to have hit Earth between 4–3.9 billion years ago. Just a hundred million years or so after that, the first signs of life appear in the rocks. This is an incredibly short time, considering the increase in complexity involved in progressing from simple organic compounds like amino acids and sugars, via replicating molecules and self-sustaining metabolisms, to fully integrated cells drawing energy from photosynthesis or inorganic respiration. Perhaps suspiciously short.

Maybe life didn't evolve that quickly at all – the first evidence of life is highly controversial and may well be a misinterpretation; the first widely accepted signs of life do not appear for another three hundred million years. (We'll look at this debate a little later.) One possibility is that the final planet-sterilising impact didn't occur so recently, giving life much longer to get going. Or, the last event may not have killed off all life but left either a few survivors clinging on in protected refuges or perhaps just some complex organic molecules, so giving the next genesis a head start. The theory of panspermia, discussed in Chapter 2, presents a third possible explanation for the rapid appearance of life as soon as the planet became fertile: perhaps cells didn't evolve quickly on Earth after the last impact but arrived, fully-formed, in a meteorite. It is not thought likely that life can voyage between different star systems, certainly not in anything larger than a speck of dust, so any such life-bearing meteorite must have originated within our own solar system, possibly on Mars. Mars is a smaller target than Earth and so would both have received fewer strikes during the bombardment and have cooled down more quickly afterwards. It is likely that Mars, if it ever were habitable, was evolving life long before Earth. These cells could have transferred to Earth in a stream of meteorites but would only have been able permanently to colonise Earth once it had become receptive, after the last mega-impact. There is at least the possibility, that humans, along with all terrestrial life, are Martian. What is distinctly more probable, considering the current lack of proof of any – present or past – life on Mars, is that Mother Earth reseeded herself. Life may have emerged in a gap between mega-impacts, been ejected off the surface within many bits of rock by a sterilising event, fallen back to Earth after it had recovered and then repopulated the planet.

However, it may be that life on Earth really did arise that quickly. Some researchers claim that the genesis of life from a pool of simple organic building blocks could occur in as little as twenty million years. Perhaps, given a broth of liquid water and organic molecules, systems that can rapidly develop into replicating molecules are all but inevitable. Many simple self-organising systems are known; once information storage and transmission have been achieved, natural selection comes into play, gradually producing fitter and more complex organisms. Such a quick start for terrestrial life is very encouraging for the prospects of it arising on other worlds, especially given the prevalence of organic molecules in nebulae throughout the galaxy.

Unfortunately, exactly what happened on Earth between the cease-fire of the heavy bombardment and the emergence of cells is one of its greatest mysteries. Prebiotic chemistry leaves no fossil record, making it very difficult to piece together the chain of events. However, it is possible to re-create the theroretical conditions in the lab, testing different environments to see which produce the best results and to consider what processes must have occurred in the emergence of life. It is to this intricate detective work that we shall now turn our attention, before moving on to the first signs of cellular life on our planet.

Prebiotic chemistry

Cells, essentially, have two subsystems. One stores the instructions on how to build the cell and either copies this information wholesale when dividing or one gene at a time when particular new components are needed. The other is the cell's metabolism, the intricate network of biochemical reactions that provide not only energy but also a rich inventory of the chemicals needed to build all the parts of the cell and copy the genetic information. Nowadays, these two functions are inextricably intertwined but at some point at the dawn of life they must have been independent. How they became linked is something we'll see shortly but first, let's consider how the organic molecules that make up both metabolism and information storage polymers might have been produced. In Chapter 3, I described how simple organic molecules have been detected in the interstellar nebulae, the clouds that will, one day, collapse to produce new stars and planets. Among the substances known to be present, hydrogen cyanide and formaldehyde are particularly important.

Hydrogen cyanide was used as a chemical weapon during the First World War but its activities on the primordial Earth would have been much more beneficial. When dissolved in water, it reacts to produce a diversity of essential biological molecules, such as urea, amino acids and the nucleic acid bases while formaldehyde polymerises, in slightly alkaline water, to form a range of different sugars. Amino acids, sugars, nucleic acid bases and fatty acids have all been found inside carbonaceous meteorites. This cornucopia of preformed building blocks would have rained from the sky during the heavy bombardment mixing, in the warm seas and lakes of the early Earth, into a lush organic soup. Even if the majority of these molecules were destroyed, their basic raw materials, carbon and water, were still vital for the emergence of life.

A second way to produce organic substances may have operated on the Earth after the deliveries of the volatile chemicals. The young planet had a reducing atmosphere. Oxygen was not produced until much later (by photosynthesis); the main constituents of the air were reduced gases, such as methane, carbon monoxide, ammonia and hydrogen. UV light from the Sun, as well as frequent lightning flashes, would have provided the energy for these gases to react together. Larger organic molecules would have been produced, gently settling out of the atmosphere and collecting in pools, just as those delivered from space might have done. Further chemical processing would have occurred in this warm water, producing what has been called the *primordial soup* or *Darwin's pond*. One famous investigation, the *Urey-Miller experiment*, carried out in the 1950s, simulated the atmosphere and warm oceans of the early Earth. Water, and a mixture of the reduced gases methane, ammonia and hydrogen, were sealed in a glass flask. The water was evaporated and taken through a tube, past a discharging spark which reproduced the effects of lightning. The vapours were recondensed and dripped back into the flask to complete the water cycle. After a week, the chemical contents of the water were analysed.

Urey and Miller found an impressive yield of amino acids in the bottom of their flask; similar experiments since then have produced sugars and other organics. For a long time, this classic piece of research was taken as convincing evidence that the origin of life had its roots in chemical synthesis in a reducing atmosphere. However, there is now some doubt as to whether the Urey-Miller experiment really reproduces the conditions of primordial Earth. First, the lightning is unlikely to have been quite as vigorous as they thought and second, complex molecules may have broken apart as they drifted down from the clouds. More important, we now believe that the early atmosphere was much less reduced. Widespread volcanism would have spewed out much more carbon dioxide than methane and both methane and ammonia are destroyed by UV light. The atmosphere was therefore much more likely to have been mainly carbon dioxide and nitrogen, with mere traces of methane and hydrogen. When the Urey-Miller experiment is repeated using this mix, the yield is practically nonexistent – the synthesis of organic substances requires a highly reducing environment.

A third mechanism for organic synthesis is currently gaining much support and also includes a plausible location for the origin of life: the hydrothermal vents strung along the tectonic spreading regions. These outlets release a rich supply of reduced chemicals and would have been much more common on the hot young Earth. The theory explains how their inorganic reducing power might have fixed carbon into organic building blocks. The proposed redox reaction is between iron sulphide and hydrogen sulphide (the gas that gives rotten eggs their disgusting smell), producing hydrogen and iron disulphide. (Iron disulphide, or pyrites, was commonly known to hapless nineteenth century prospectors as 'fool's gold'.) The iron is oxidised, releasing electrons that can be donated to the copious carbon dioxide bubbling out into the ocean water. Reduced organic molecules are created, effectively fixing the carbon just as photosynthesis does using light. Not only that, but the growing crystals of iron pyrite are also very good at binding to the newly reduced organic molecules, holding them to a solid surface. This acts as a catalyst, increasing the reaction rate between different simple organic substances to produce larger molecules. The walls of the black smoker chimneys are also very porous, providing lots of tiny pockets where organic substances can become concentrated. Within these pockets, small, self-sustaining, cycles of reactions might have set up, with fresh reactants diffusing in from the vent plume and waste products washing away. These rocky reaction chambers could have formed the template for the first cells, with a membrane of fatty acids replacing the open wall.

Experiments simulating the conditions found in hydrothermal vents have re-created a great deal of interesting chemistry, producing long-chained carbon molecules and amino acids that polymerise into short protein strings. Some researchers believe that the very first metabolic networks appeared in the dark depths around hydrothermal vents, rather than in the bright primordial ponds on the surface. The role of iron in this hypothetical genesis is particularly interesting. Iron, along with other metals such as nickel, copper and zinc, can readily switch its redox state, either accepting or donating electrons. This makes these metals very useful for facilitating redox reactions in metabolism; more than a third of enzymes contain at least one metal atom. This means there could be a direct link between the metal ions spewing out of hydro-thermal vents and the complex enzymes used by life today.

As with all other ideas on the origins of life, the accuracy of the black smoker theory is difficult to determine. Not only has all the original oceanic crust long since been destroyed by tectonic recy-cling but lab experiments have been slow to demonstrate that metabolic networks can arise in vent environments. Some researchers think that the vents were a net remover of accumulated organic substances, not a source. The entire ocean is circulated through hydrothermal vents every ten million years or so, destroy-ing any dissolved organic substances in the superheated water.

The first organic substances probably arose from a combination of sources; from above, below and beyond (from lightning and UV light in a reducing atmosphere, from hydrothermal vents and by delivery from outer space). There is also a range of theories of the order of events of biogenesis. Some researchers support the sequence of membrane bubble, proteins, genetics, whereas others favour the primacy of genetic information storage. The process that I shall argue for is that metabolism predates genetics, with pro-teins and an enclosing membrane being incorporated along the way.

A consensus about how the first metabolic networks were built up is emerging. The key is a set of *autocatalytic* reactions. In such reactions, each step is catalysed by one of the other products of the network. In a large enough pool of chemicals, engaging in differ-ent reactions, there is a good chance some will catalyse each other. These products thus become more common and, as a greater diversity of reactants is generated, there are more potential catalysts to act on the others. This process accelerates as an increas-ing number of molecules are produced and the network of catalytic interactions becomes denser, until a single, self-sustaining, network spontaneously emerges. This network acts to produce more and more of the compounds in the cycle and, over time, more reactions become involved, enlarging the circuit or adding branching points. Catalytic surfaces, such as inorganic compounds or clays, are probably also involved.

Some sections of metabolism are common to all life and must have appeared very early in prebiotic chemistry. As these ancient biochemical networks became increasingly complex, they began producing small organic substances and polymers that had not previously existed and that weren't produced by UV light acting on vapours or inorganic reactions in black smokers. Although it is far from certain whether metabolism pre-dates genetics or not, the complexity of modern information storage molecules, RNA and DNA, strongly indicates that primitive biochemical networks must have existed first, to provide both the necessary components of, and the energy to drive, the synthesis of any polymer complex enough to store and copy information reliably. The next part of the story is the emergence of these replicating polymers.

RNA world

Until the 1980s, there was thought to be a deep paradox in the emergence of life. DNA codes the genetic information that is translated into proteins but this process requires enzymes, energy and precursor molecules, supplied by a metabolic network. Modern metabolic networks are fabulously elaborate and cannot operate without the catalytic action of enzymes, themselves coded by DNA. Therefore, DNA cannot function without metabolism and metabolism cannot function without DNA. There is a deep chicken-and-egg problem at the beginning of life.

This paradox was resolved by the discovery that RNA molecules, already known to store information within the cell, also had catalytic properties – they could assist certain chemical reactions, like enzymes. These *ribozymes* are a bridge between the realms of biochemical networks and replicating information polymers. For a time, they are thought to have dominated the planet, a period referred to as the *RNA world*.

The first ribozymes were probably no more than short chains of nucleic acids, folded into a limited three-dimensional structure that allowed them weakly to accelerate biochemical reactions. A variety of different ribozymes co-existed, some catalysing more useful reactions than others within the primordial metabolic networks. We do not yet know the full range of possible reactions of ribozymes but it probably includes a variety of redox reactions and perhaps the synthesis of sugar polymers. The current research goal is to demonstrate experimentally that they can catalyse the production of their own building blocks, nucleotides, or harness redox reactions to link amino acids into a protein. However, at some stage, ribozymes must have emerged that were able to catalyse the most important reaction of all: to synthesise other strands in their own likeness and thus become capable of self-replication. Once this trick was learned, as well as driving metabolism, RNA could reproduce itself and the powerful engine of natural selection could start. A fierce rivalry would have existed between different varieties, with the less effective ribozymes eventually being outcompeted. This very first occurrence of evolution gradually produced ribozymes able to replicate faster or more accurately or co-operate for their mutual benefit. Unions of different ribozymes would have formed, one perhaps specialising in producing nucleotides, another churning out fatty acids to expand the membrane containing them and a third catalysing the replication of itself and the other two. When the fatty-acid bubble swells to an unstable size it splits into two: the basis of cell division. RNA replicators would be prone to errors, introducing mutations into the ribozymes they were copying. These new variants might catalyse different reaction steps, improving the range of precursors the system could use.

A second revolution occurred: the appearance of RNA molecules containing a code that did not describe how to build themselves but how to build a protein. Other ribozymes read this code and strung together the required amino acids. These genetically-encoded proteins came under evolutionary pressure, adapting them into increasingly effective catalysts. Thus, the RNA world developed a more productive workforce, a set of enzymes able to take over the task of running the metabolism and replicating the genes, while RNA remained responsible for holding the genetic information. At some stage in the growing complexity of these proto-cells, they stepped over the ill-defined line dividing the living and non-living. Primitive, membrane-bound, systems of genetics and metabolism became more like a bacterium than a simple soup of interacting chemicals.

During this upgrade from the original scheme of ribozymes to the new, high-tech proteins, the evolving genetic system hit a wall. These first, mini-genes, were perhaps 100 bases long but could grow no longer. RNA is not particularly stable; although it is complex enough to code information, its capacity is severely limited and it is not a very good long-term store; a hardier molecule was needed.

The protein that ushered in this new storage format is *reverse transcriptase*, an enzyme capable of transforming the information contained in RNA into a DNA molecule. This function is used by retroviruses, such as HIV, and is a central process in genetic engineering. Using reverse transcriptase, RNA genes could be converted into a much more robust and stable material, like a scribe transferring his work from wax tablets to durable parchment. Modern genes are many thousands of bases long (the human genome has around three billion bases, arranged on twenty-three paired DNA strands). RNA lost its role as the primary data-storage medium within the cell but did not become totally subservient; RNA remains the linchpin of life, holding together the different subsystems of a cell. Messenger RNA acts as a temporary copy of the genetic information stored in the DNA library and ferries it to the ribosome (itself based on RNA), which churns out the specified protein by linking together spare amino acids, each attached to a transfer RNA molecule. The ribosome is effectively a giant ribozyme, decorated by attendant proteins, a relic of an archaic world that has taken on critical duties within the cell. The RNA world is far from dead; RNA viruses still vastly outnumber cellular life. In some ways, the interior of a modern cell is the persistence of conditions prevalent in the RNA world and RNA viruses are merely continuing an ancient way of life.

The evolving system of the RNA world is theoretical; its existence so far back in time can never be proven, only experimentally demonstrated. There is also an increasing suspicion that RNA may not have been the first molecule capable of self-replication. It is difficult to explain the synthesis of RNA using prebiotic chemistry: formation of the bases is apparently easy, created from inorganic hydrogen cyanide from meteorites but the manufacture of ribose is much harder. Formaldehyde, also present in meteorites, is known to produce ribose when it reacts in alkaline water. But this non-specific process produces ribose in low concentrations and as just one of a host of irrelevant sugars.

One possible resolution is that ribose never was created by abiotic chemistry. We've already touched on the idea that metabolism probably came before genetics; ribose may not need to be formed abiotically but could be the product of an advancing biochemical network. If this is the case, it is entirely possible that simpler information-carrying molecules may have emerged first. TNA, the polymer built from a four-carbon sugar that I discussed in Chapter 1, is one possibility. There may even have been a succession of genetic systems, each out-competing a more primitive rival, until the hegemony of the DNA-protein world took over in the first cells, with the previous reigning champion, RNA, clinging on as a go-between.

Cellular life

All life on Earth uses a DNA/RNA genetic system and an array of proteins oversees its metabolism. The *last universal common ancestor* (LUCA) must have possessed these attributes before the archaeal and bacterial domains diverged. The LUCA is unlikely to have been a single cell-type; more likely a community of organisms engaged in an orgy of gene exchange, a distributed *progenote collective*. As different life forms evolved and their cellular systems became idiosyncratic and less flexible, the opportunities for gene sharing lessened. The fundamental protein translation machinery 'solidified' first; metabolic enzymes are still being passed between distant relatives. Today, there are two prokaryotic cell types – archaea and bacteria – but it is possible that other, fundamentally different, life forms split from the common ancestor, evolved, exchanged genes, then succumbed to extinction. This might explain the curious presence, in eukaryotes, of some genes that don't seem to resemble anything found in either archaea or bacteria.

One of the biggest questions about the very first cells was whether they were autotrophs or heterotrophs. Were they self-reliant, able to extract energy and fix carbon from inorganic reactions (autotrophs) or did they need to consume preformed organic molecules from their surroundings (heterotrophs)?

If the primordial atmosphere was sufficiently reducing, abiotic reactions, driven by UV, lightning or black smokers, might have provided the first cells with a ready feast of complex organic substances. As the biosphere grew, this food supply would have gradually dwindled, forcing the hungry cells to evolve methods of generating the substances they needed for themselves; moving from heterotrophy to autotrophy. In a less reducing atmosphere, autotrophy might have developed first, with heterotrophy a later adaptation to living off the efforts of autotrophs.

One way of settling the argument is to look at the tree of life, which plots the relatedness of different cells. The root of this tree, the current organism that is most closely related to all other organisms, would give us some indication of what the universal ancestor was like. Unfortunately, drawing such a wide-ranging family tree is very difficult; depending on what assumptions are made and what statistics are used, the root moves about. For now, it is impossible for us to tell whether the LUCA was autotrophic or heterotrophic. However, one pattern is common to the slightly different trees that have been constructed. Most show an archaeal hyperthermophile in the middle, suggesting that the mother of all life on Earth lived in very hot water. This lends weight to the theory that life emerged in the chimneys of deep sea vents but another interpretation is equally plausible. Towards the end of the heavy bombardment, impacts large enough to completely boil away all water and sterilise the Earth had stopped but smaller impacts, dumping enough energy to heat the oceans substantially would have continued. Perhaps it was not that the first life was thermophilic but simply that the only cells that survived to the end of the heavy bombardment were those that could tolerate these temperature bottlenecks well enough to repopulate the world. Not so much 'thermophilic Eden' as 'thermophilic Noah'.

First signs of life

Very few rocks date back to the earliest period of our planet's history. Even if an unbroken geological record existed, the action of prebiotic chemistry would have left no trace. The first signs of cells in the geological record are highly ambiguous, reflecting a central problem of astrobiology – how can life be unmistakeably identified and distinguished from similar abiotic processes? Three main lines of evidence have been used to date the first appearance of terrestrial life.

The earliest evidence of biological action is argued to be in a patch of 3.8 billion-year-old rocks in Greenland. These are some of the most ancient rocks on the surface of the Earth and have been extensively modified by heat (the process of *metamorphosis*) over the aeons. Some researchers claim that the graphite trapped within their mineral grains is rich in the lighter isotope of carbon, a trademark of biological processes. (Isotopes are forms of the same element that have different numbers of neutrons in their nucleus. Carbon has several isotopes, including carbon-12, with six neutrons and carbon-13, with seven.) ^{12}C and ^{13}C are naturally found in the environment at a known ratio: many biological processes favour the lighter isotope, because it is more easily captured and released from chemical bonds, which distorts this ratio. For example, rubisco, the enzyme that fixes carbon dioxide during photosynthesis, produces organic compounds with around two per cent more ^{12}C. The researchers claim that just such a change in the isotope ratio is present in the Greenland rock and must have been the result of biological action while the rock was still sedimentary and before it underwent metamorphism. This claim is very hotly debated; its critics point out that certain abiotic reactions also cause distortion of the $^{12}C:^{13}C$ ratio, that the rock could not originally have been sedimentary and even that no carbon is present in them at all. There is an increasing rejection of such a weakly supported, indirect and extremely early dating of life.

The second line of evidence for a very early emergence of life rests on a layer of rock in the Warrawoona outcrops in Western Australia. Their supporters claim that these rocks, 3.45 billion years old, contain microscopic features suggestive of cells – in fact of eleven species of fossilised bacteria. The size and shape of some of these suggests they may be *cyanobacteria*, the only group of prokaryotes that can perform oxygen-releasing photosynthesis. This, in itself, is an extremely troublesome claim, as it predates, by about a billion years, many other signs of cyanobacteria and evidence for an oxygen-bearing atmosphere. It is hotly debated whether they are cells at all, as similar forms can be produced by purely geological processes, such as the hydrothermal crystallisation of minerals.

The third class of evidence involves the dome- or column-shaped layered rock formations *stromatolites*. Today, such structures are created by mats of photosynthetic and heterotrophic bacteria living in warm, shallow water. Mucus excreted by the cells traps drifting sediment and binds it into a covering layer. The bacteria recolonise the new top surface open to the sun, and so over time these *biofilms* build into large structures of layered sedimentary rock, as seen in Figure 8. Similar-looking structures have been found in 3.5 billion-year-old rock, in the Warrawoona area, and have been cited as evidence of bacterial action, including photosynthetic cyanobacteria, at this very early date. No fossils of cells are found in these ancient structures and the curved layers could simply have been produced by the deformation of abiotic

Figure 8 Modern stromatolites in the shallow water of Shark Bay, Western Australia.

sediments. Stromatolites were undoubtedly one of the first mani-
festations of life: the debate, as with the other lines of evidence, is
about just how soon the first signs are unmistakeably the work of
biology.

Although the exact timing of the emergence of life is very controversial, the evidence is overwhelming by about three billion
years ago. By this time, prokaryotes had become widespread,
dotted around deep and shallow water hydrothermal habitats,
drifting near the surface of the open ocean and lining the coastal
fringes, forming ecosystems of photosynthesising autotrophs supporting grazing heterotrophs.

We briefly considered photosynthesis in Chapter 1 but it has
played such a pivotal role in the Earth's evolutionary history that it
is time to return to it in detail.

Photosynthesis and the oxygen revolution

Autotrophy enables cells to be independent of organic molecules
for their energy or carbon requirements but chemosynthesis still
restricts life to small regions with steep redox gradients – such as
around black smokers or in hydrothermal pools. The development
of photosynthesis – surviving on light – liberated cells from these
locations and left them free to colonise the whole planet.

The chlorophylls, which form the basis of photosynthesis, are
believed originally to have evolved to protect the proteins and
DNA of early cells against the harsh UV radiation streaming down
on to the surface. Presently, UV is absorbed by the ozone layer, produced by the action of sunlight on the oxygen in the atmosphere
but three and a half billion years ago, the atmosphere was almost
devoid of oxygen and there was little protection from UV radiation; surface life therefore evolved molecules able to absorb this
light. Once these have absorbed a particle of light, they must

dissipate its energy before they can absorb more. One way of doing this is to pass excited electrons on to neighbouring proteins. The first photosynthetic organisms used this 'waste' electron energy to fix carbon and drive useful redox reactions. However, there is a constraint to the derivation of energy in this way: the cell needs a steady supply of replacement electrons, meaning it must be near a source of reducing power. Some bacteria gain these electrons by oxidising hydrogen sulphide, a chemical abundant in hydrothermal pools but to be fully autonomous a cell must cut this final tie.

The solution that nature found was to take electrons from water molecules. Thus, photosynthesis could provide a renewable energy source and release life from its dependence on reduced ions in the environment. This development required a new, supplementary, photosystem, as it takes a great deal more power to split water but it has the benefit of providing enormous amounts of energy for the cell.

The pollution created by this advanced form of photosynthesis is responsible for one of the greatest ecological disruptions in the history of life. The splitting of water releases an extremely poisonous gas that the photosynthesising cells excrete into the air. This toxic chemical forms highly reactive products, such as peroxide, that ravage the complex organic molecules of life. It is, of course, oxygen and the vast amount of it in today's atmosphere is there as a direct result of photosynthesis. The only organisms capable of splitting water in this way are cyanobacteria and the plant cells that formed an endosymbiotic relationship with them.

Oxygenic photosynthesis could have evolved as early as 3.5 billion years ago but the first evidence of rising oxygen levels in the atmosphere does not appear until 2.5 billion years ago. The atmosphere and water became steadily more oxidising and the reduced minerals exposed to it were oxidised. Rocks dating from the period show thick bands of deposited iron oxide – the surface of the planet was literally rusting. Once the Earth's crust had soaked up all the oxygen it could, its atmospheric level started to rise

rapidly. Life had previously been anaerobic (able to run its metabolism without needing oxygen as a final electron acceptor):
indeed, much life still finds even trace amounts of oxygen
extremely toxic and is confined to places, such as deep water or
underground, where it cannot reach. Aerobic life survives because
it developed a suite of mechanisms to cope with oxygen's destructive effects. By 2.2 billion years ago, the atmospheric concentration
of oxygen had climbed to about one per cent of its present value
and suddenly something happened to life on Earth. Previously,
cells (including the photosynthetic cells that were generating it in
the first place) had simply limited the damage that the products of
oxygen reactions could wreak. One method is to produce scav-
enger molecules that combine with and neutralise the reactive
oxygen chemicals before they could attack crucial polymers. But
some organisms took a new tack and turned this immense oxidising power to their advantage. As I described in the section on respiration in Chapter 1, oxygen makes a very good final electron
acceptor at the bottom of the transport chain. By exploiting
oxygen in this way, aerobic cells were able to extract far more energy
from the organic molecules by breaking them down even further.

The rising atmospheric oxygen levels had another important
effect on surface life. Solar UV light splits apart oxygen molecules
high in the atmosphere, producing ozone, which is itself a strong
absorber of UV. So as long as there is an appreciable presence of
oxygen in the air, the action of sunlight naturally builds up a strong
ozone shield, limiting the amount of UV radiation that can reach
the ground. On the early Earth, the levels of DNA-damaging UV
radiation are estimated to have been more than forty times more
intense than today.

The importance of photosynthesis cannot be overstated. High
concentrations of reduced inorganic compounds are a relatively
limited resource and this probably constrained the spread of early
life. The evolution of photosynthesis provided light-harvesting
cells, and the ecosystems they supported, with an effectively

unlimited source of energy. The oxygenic process additionally produced oxygen, allowing more efficient burning of nutrients and creating an ozone shield against UV radiation. Photosynthesis allows cells to rebuild themselves using simply carbon dioxide, water and the energy of sunlight. These simple building blocks are likely to be common on other terrestrial planets; so photosynthesis could provide metabolic energy to alien ecosystems across the galaxy. Porphyrins, the basis of chlorophyll, haemoglobin and similar metal-containing molecules, are synthesised extra-terrestrially; so chlorophyll may be widespread on alien worlds. In later chapters we'll look at the plausibility of harvesting the light of other stars and how astronomers might detect the tell-tale signs of photosynthesis from light years away.

Eukaryotes and multi-cellular organisms

The earliest eukaryotes appeared around two billion years ago, although exact dating is difficult because the first cells were as small as archaea or bacteria and had no cell wall to leave a recognisable fossil. Their evolution intriguingly corresponds with rising oxygen levels. This is probably no coincidence and the evolution of the nucleus may have been driven by the need to protect the fragile DNA from oxidation. Dissolved nutrient levels were also low at this period; endosymbiosis, in which mitochondria and chloroplasts were taken in as organelles, may have improved energy efficiency and recycling between previously free-living organisms. Many eukaryotes, such as our own human cells, are aerobic; an oxygen-laden atmosphere is crucial to their existence. In this sense, oxygen is a double-edged sword for complex eukaryotes – they cannot survive without it but it is also powerfully destructive if the cells do not handle it properly. Many of the more risky energetic reactions are carried out by mitochondria which, as I described in Chapter 1, are believed once to have been free-living bacteria which took up their lodging in the developing eukaryotic cell. Chloroplasts are descended from cyanobacteria, the only prokaryotic group capable of oxygenic photosynthesis.

As well as their sophisticated interior organisation, improved DNA capacity and genetic control, eukaryotes have one other enormous advantage over prokaryotes – an evolutionary leap they made about a billion years ago. Eukaryoates are exceptionally good at getting together; building much larger organisms from congregations of single cells. The adult human body, for example, is a colossal colony of some hundred million million cells, sacrificing their individuality to the health of the whole organism. There are over two hundred specialised types of cell, each expressing a subset of the body's complement of several hundred thousand different proteins. And this exquisite complexity self-organises from a single fertilised egg cell. Multi-cellularity has independently evolved many times, including among the fungi and microscopic protists, but animals and plants are particularly accomplished at it. Although prokaryotes often stick together into microbial mats, such as stromatolites or the plaque that forms on your teeth, true multi-cellularity is distinct from simple colonial living. Even in a primitive animal, the cells are differentiated into a great diversity, each with a specific function and anchored on to a sturdy extra-cellular matrix, communicating with each other and surrendering their reproductive rights to a few germ-line cells. A few examples of multi-cellular organisms exist in the bacterial world but they are comparatively simple and mostly temporary arrangements. Myxobacteria, for example, release a signal when nutrients begin running out and swarm together into a structured slime, which produces a single upright stalk of co-operating bacteria that eventually releases a puff of reproductive spores. But nowhere on Earth are there fronds of bacterial seaweed or archaeal jellyfish.

The lack of widespread multi-cellularity among prokaryotes is surprising, considering its clear advantages. Larger organisms are safer from predators, more effective at nutrient uptake and storage, can better regulate their internal environment and the division of labour can make the whole organism more versatile. Perhaps prokaryotes are simply incapable of the sophistication of communication or genetic control required. However, they remain the most successful organisms on the planet, perfectly adapted to rapid reproduction and capable of exploiting an enormous variety of energy sources.

The Cambrian Explosion and 'Snowball Earth'

The earliest complex animals are dated to around 750 million years ago by some lines of genetic evidence but the first fossils do not appear until about 600 million years ago. Complexity demanded a number of crucial developments, including the protein *collagen* – a component of the extracellular matrix that holds animal cells together – and *homeobox* genes, a set of switches that establish the features of a body plan.

Around 540 million years ago, the level of oxygen in the atmospheric was about ten per cent of its current value. This date is coeval with another revolution for life on Earth when, quite suddenly, a spectacular diversification of marine animals appears in the fossil record. Every basic body form we see today, such as molluscs (like mussels and snails), arthropods (like lobsters and insects) and chordates (like humans and fish), first appeared in this period. The fossils also document many weird and wonderful forms that have not survived to the present day, such as the armoured predator with a pair of jagged tentacles and rows of swimming fins, depicted in Figure 9.

The cause of this apparent flurry of evolutionary innovation, the *Cambrian Explosion*, is hotly debated. Some scientists argue that it does not represent a sudden biological event but rather that

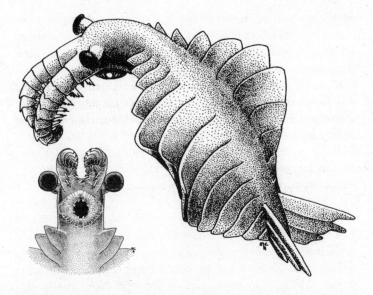

Figure 9 Artist's impression of Anomalocaris, a long-extinct bizarre animal life-form appearing after the Cambrian Explosion.

fossilisation conditions changed and only then were pre-existing species preserved. Others claim that a specific incident sparked an evolutionary race between competing organisms. For example, the evolution of the eye would have allowed active hunting, triggering an arms race of adaptations in predators and prey.

Another explanation is particularly relevant to astrobiology and the implication of the stability of planetary climates. In Chapter 3, I described how planets within the habitable zone of their star tread a thin line between opposing disasters: rising temperatures can set off a runaway greenhouse effect, while runaway glaciation results in plummeting temperatures and a global covering of ice. Earth's current climate seems to be unusually cool; for ninety per cent of its history there was no ice cap on the poles. But records of previous climate, held in rocks, also show that on several occasions between 750 and 600 million years ago a deep glaciation befell the planet. Not only did these episodes collapse the surface ecosystems, they also came close to wiping out eukaryotes altogether.

Once the cooling started, the growing regions of white ice reflected more and more of the Sun's warmth, creating ever-colder conditions. The polar ice caps expanded, until they almost touched at the equator and may even have blanketed the entire surface in a thick ice sheet. The seas of this *Snowball Earth* were sealed, blocking out the sunlight and causing a precipitous drop in photosynthesis. Deprived of its source of oxygen, the seas turned anoxic, for the first time in over a billion years. The Earth effectively suffocated and great numbers of organisms would have become extinct. The fact that we are still here implies that some prokaryotes and our eukaryotic ancestors clung on, surviving in havens of meltwater or thin ice on the super-continent or holes in the glaciers around volcanic islands. Some researchers are dubious that the surface ever completely froze and argue for more of a 'Slushball Earth', with pockets of clear water.

Either way, each of these epic Ice Ages lasted several million years, ended only by the saving grace of our planet's active volcanism. With erosion of silicate rocks almost non-existent, carbon dioxide accumulated in the air to as much as 350 times its current level. The powerful greenhouse effect created by such a well-insulated atmosphere finally began to thaw out the planet. During this recovery period, the weather would have been absolutely frenzied, as warm air from the tropics expanded into the still-frozen wastelands around the poles. Giant ripples in sedimentary rocks from this time suggest long-running storms assaulted the seas with wind speeds of over seventy kilometres per hour. The climate eventually settled into equilibrium and life repopulated its former niches. The theory is that the release from such a harsh environment provided the evolutionary prompt for complex animals, leading to the Cambrian Explosion.

As this journey through time approaches the modern day, we see multi-cellular plants and animals joining the prokaryotes in colonising the land, the once-barren continents flushing green with vegetation, as plants evolved flattened leaves, rigid wood and finally flowers and fruit. Animals similarly develop their own adaptations, including yolky eggs, flight, warm-bloodied physiology and now self-conscious intelligence as the continents slide into their familiar positions.

Extinction is as much a natural part of evolution as innovation and the emergence of new forms. The constant background level of extinction is like the soft crackle of static on an untuned radio but every now and then a great pulse of death strikes the entire planet. Since the Cambrian Explosion, there have been five devastating mass extinctions, each wiping out a great fraction of marine- and land-based life. Such catastrophes might be linked to events in the galaxy around us, from nearby exploding stars to dense clouds of space dust but these cosmic processes are essential for life, creating vital elements and organic molecules.

In terms of astrobiology, multi-cellular life, especially complex, land-living, animal life, is possibly something of an anomaly. The vast majority of the galaxy is probably only suitable for microbial life, hardy bacteria-like organisms able to survive in relatively harsh environments. Earth's extremophiles tolerate boiling acid, saturated salt or freezing parched conditions and chemoautotrophs can subsist on nothing more than inorganic gases bubbling through deep rock. Higher forms of life are much more sensitive to environmental upsets, require longer periods of stability to evolve and need established ecosystems to provide their greater energy demands. Any extra-terrestrial life we are ever likely to encounter will probably be analogous to our prokaryotes, with many fewer inhabited worlds having progressed to the eukaryotic stage, let alone multi-cellular animals or land ecosystems. The galaxy may well be teeming with different life forms but we'll need a microscope to see them. Certainly within our own cosmic back garden, the solar system, few researchers hope for anything more complex than unicellular prokaryotes.

We'll first turn our gaze towards our neighbouring planet, Mars, before considering Venus and then heading further out to the moons Europa and Titan.

5

Mars

Mars, the fourth rock from the Sun, has received more attention about the prospects of extra-terrestrial life than any other world in the solar system. The presence of life on Mars has been widely accepted for long periods; even into the 1960s, textbooks explained the fluctuations of the dark patches visible on its surface as seasonal variation in vegetation. Perhaps the most notorious proponent of a living Mars was Percival Lowell. During the early twentieth century, this American astronomer heavily publicised his observations of an extensive network of lines criss-crossing the Martian surface. Sightings of *canali* on Mars were first reported, as possible narrow seas, by the respected Italian astronomer, Giovanni Schiaparelli. But Lowell pushed the interpretation further, arguing the *canali* were artificially constructed by a civilisation far more advanced than any on Earth, desperately trying to survive on a dying planet by transporting meltwater from the poles to the cities and farms around the equator. Such sightings and the exuberant claims based on them gradually became discredited as tricks of the eye; remaining hopes of Martian civilisations were finally dashed by a string of robotic probes.

The first explorer to reach Martian orbit was Mariner 9 in 1971. For several weeks after arrival, its view was completely obscured by a worldwide dust storm. As the dust settled, the probe's cameras began sending back shots of the planet's surface. There was no trace whatsoever of Lowell's network of canals but the view was no less stunning for that. The eager scientists could make out a cluster of huge volcanoes towering above the wispy clouds and a great gash along the planet's equator, named Valles

Marineris in honour of the probe. The entire surface of the planet has now been laser-mapped to an accuracy of metres; we know the landscape of this alien world better than that of our own.

We see signs on Mars of many of the processes rife on Earth. The wind whips up mini-tornado 'dust devils' and full-blown dust storms. Evidence of fluid action is widespread: valleys coursing through the highlands, eroded headlands, partially-erased craters and extensive deltas of deposited sediment. The younger Mars was much more volcanic and there is tentative evidence that limited plate tectonics may have occurred very early in Martian history. These geological processes are thought to be crucial to terrestrial life so let's take a short look at the features of an energetic Mars.

The geography of Mars is the most extreme of any planet, with a thirty-kilometre height difference between the highest peak and deepest basin. The Tharsis bulge is a colossal swelling of the Martian crust; because the planet has never had any substantial plate tectonics this hot spot of magma has remained beneath the same location on the surface, allowing very long-lived volcanoes to grow to a huge size. A chain of three giant volcanoes straddles the equator, each utterly dwarfing any mountain on Earth. Just to the north-west of these lies the highest mountain in the entire solar system. Olympus Mons towers twenty-five kilometres into the Martian sky, almost three times higher than Mount Everest, poking above most of Mars' atmosphere. Valles Marineris runs east-wards for four thousand kilometres away from the Tharsis bulge, cutting seven kilometres deep into the crust in places, making the Grand Canyon look like a superficial scratch. It is one of the few actual features on Mars that can be identified as one of Lowell's canals. The lowest point on Mars is the enormous Hellas impact basin, its outer rim marked by a ring of mountains over four thousand kilometres across. This crater was gouged out during the heavy bombardment, with much of the enormous volume of excavated rock contributing to the surrounding high ground. Perhaps the most noticeable geographical feature of Mars is the stark difference between terrain in the northern and southern hemispheres. The mountainous highlands in the south are very old and heavily cratered from the bombardment, whereas the northern basin is mostly flat and smooth, so must have been completely resurfaced relatively recently. The importance of this will become apparent shortly.

Water, water, everywhere

For astrobiologists, the most important question is whether Mars has liquid water. The early results provided by Mariner were not encouraging. Later probes confirmed that Mars does not have a thick atmosphere – the surface pressure is less than one per cent that at sea level on Earth. Although it is almost entirely carbon dioxide, such a thin atmospheric blanket provides only a pitiful greenhouse effect and so Mars is bitterly cold. The ground on a summer's midday on the equator can approach room temperature but most of the planet is well below 0 °C for much of the year, with the polar caps dipping to a chilly −140 °C. This combination of very low air pressure and freezing temperatures means that, generally, liquid water is not possible on the surface of Mars (see Figure 5 on p. 61). Martian conditions are below the triple point of water and even if ice warms in the sun it turns directly into vapour without first running as a liquid. If Lowell's canals existed they would never flow, even with extremely salty water. Any exposed ice near the equator has long since been boiled away by the Sun and the only visible surface water is around the poles.

Optimism began to return when further probes beamed back better-quality images. Intriguing landforms could be seen on the cold surface, telling the story of a very different Mars. The new photographs clearly showed wide channels running down to the northern lowlands (see Figure 10) and great networks of valleys carving meanders through the rough terrain of the highlands – looking for all the world like river courses on Earth. The prospects for flowing water on Mars, if not today, then at least at some point in its past, were much more promising. Could the conditions on the young Mars have been much more clement, with a thicker atmosphere to better insulate the planet and provide conditions for liquid water? These distinctive features seemed to say so but all was not quite what it seemed.

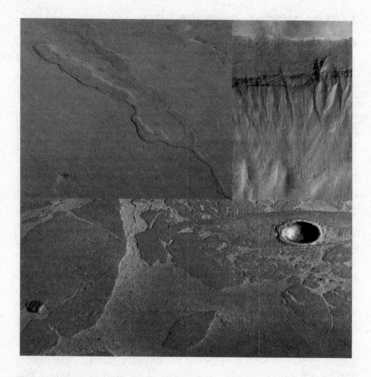

Figure 10 Signs of liquid water, ancient and modern, on Mars. Clock wise from top left: two ancient outflow channels emerging from the flanks of a volcano; young gullies running down the side of an impact crater; pack-ice-like features suggestive of a recent surface sea.

Despite looking distinctly Earth-like, there are some very unusual features of the valleys and channels on Mars. The valleys are almost only ever found in the southern highlands, suggesting that they are extremely ancient. Tributaries can be seen combining into larger courses that begin meandering downstream, eventually spilling out on to wide flat plains or flowing into impact craters. The enormous Hellas basin has a system of associated valleys, arranged radially around it like the spokes of a bicycle wheel. The suspicious thing is that the tributaries don't seem to gradually grow; they pop right out of the ground. The valleys are not thought to have been fed from falling rain, like most rivers on Earth, but to have been formed by reservoirs of underground water breaking out on to the surface.

The drainage channels are stranger yet. Their scale is absolutely staggering; far wider than any river on Earth, they are tens of kilometres across and hundreds or even thousands of kilometres long, stretching from the highlands into the northern hemisphere of the planet. Like the valley networks, these channels simply emerge from the ground. At the source are broad areas of disordered terrain, some the size of Switzerland, as if the surface had simply collapsed on itself. Many of these outflow channels run from active areas, such as the great volcano complexes of the Tharsis bulge and it is thought the overwhelming floods were released suddenly by geothermal events. Much of Mars' water may have been stored in the crust, locked in an underground layer of permafrost. This ice shell would have trapped liquid water beneath, perhaps building to great pressures until a volcanic episode created a rupture and released a torrent of meltwater. Once on the surface, dissolved carbon dioxide would instantly fizz out of solution as the pressure dropped, causing the spurting cascade to resemble the uncorking of an enormous bottle of champagne. Large outflows could also have been triggered by a nearby impact, as the shock waves shattered the permafrost seal. Some calculations show that a single torrent could unleash up to a billion cubic metres of water a second. The largest known catastrophic flooding on Earth was the filling of the Black Sea eight thousand years ago (a possible source of the story of Noah). Martian outflow channels were produced by deluges hundreds of times greater than this, the result of floods on a truly Biblical scale.

From an astrobiological point of view, these extremes of solid permafrost layers and flash surface floods are interesting, demonstrating the possibility of large amounts of water, but they do not provide an attractive location for the emergence of life. Life is thought to need sites where organic molecules can become concentrated and react together, such as warm surface pools of liquid water, or perhaps around hydrothermal vents in deeper water. Does the Martian surface show any evidence of long-lived bodies of standing water? The valley networks do often end up in basins where water could collect and some craters display neat concentric rings of sedimentation where a lake has dried out or repeatedly been topped up. Such impact crater lakes may even have rested on a hydrothermal heating system, as discussed in the extremophiles chapter. The volume of water released along the outflow channels would have prevented it from freezing quickly, even under the current climatic conditions. The northern lowlands constitute the largest drainage basin on the planet and the vast floods coursing along the channels would have spilled out across this plain. Surely this is where a sea would collect, possibly surviving for quite some time before sinking away into the ground or subliming into the air. Some researchers studying satellite images of the northern basin have claimed there is evidence of the shoreline of an ancient ocean. They point to a series of distinctive features that form a continuous line running at roughly the same height as might be expected for a sea level. The plains below this contour are particularly smooth, perhaps an even covering of sediment which settled over the floor of an ocean. Depending on the environmental conditions, the surface of this sea might have been littered with great blocks of drifting pack-ice or frozen over almost to its floor or (a likely possibility given the volume of detritus that would have been picked up by the floods) a lake of thick mud. The general feeling however is that a sea would not have been possible at all during the epoch of the flood channels. Any surface water would most likely have soaked away into the crust or escaped into the atmosphere before the next deluge arrived.

Direct evidence of liquid water

These clues I have described to the history and extent of liquid water on Mars have been based on circumstantial evidence – photographs taken from orbit. Until recently, it was not even certain whether water was the fluid that created these landforms. The valley networks, outflow channels and crater sediments could be, at least partly, explained by non-aqueous processes, such as other liquids, lava flows, wind or even a fluid-like flow of dry dust. The final resolution was not to come until 2004, when a pair of explorers landed on Mars and delivered conclusive proof of persistent liquid water on the surface. Within three weeks of each other, the twin Mars rovers *Spirit* and *Opportunity* bounced on to the Martian surface, cushioned within a cocoon of airbags. Both found evidence of liquid water but I shall focus on the more astounding discoveries of the second twin.

Opportunity's target was a flat plain in the northern hemisphere, the *Meridiani Planum*, chosen because it had been found to contain an exceptionally high level of an iron oxide mineral, haematite. Many of the processes known to form haematite require liquid water, so NASA scientists were keen to get a probe there to investigate. As soon as *Opportunity* raised its panoramic camera and looked around, the team knew they had struck gold. *Opportunity*, by sheer luck, had landed smack in the middle of a shallow impact crater, its floor covered in a fine dark soil, littered with mysterious marble-sized nuggets the scientists nicknamed 'blueberries'. And on the crater wall was exposed bedrock with a clear pattern of sedimentation layers – the perfect place to hunt for signs for liquid water. This is its story, as the scientists unravelled it.

The ground rock at Meridiani Planum is undeniably sedimentary, laid down within shallow seas created by episodic flooding. The chemical details of the rock indicate that water was definitely the liquid involved and was fairly warm, at least $-10\ °C$. But the high content of a sulphate mineral, *jarosite*, means that the water must have been quite acidic, probably due to dissolved volcanic gases. This acid water leached salts out of the ground, making the sea also very briny. Over time, the water evaporated away to leave dry salt and dust blowing across a plain. But underground, the rock remained sodden and iron compounds precipitated to produce growing lumps of haematite. These lumps were subsequently eroded out of the rock by ferocious dust storms and scattered across the ground as the blueberries seen on *Opportunity's* first day. This cycle of flooding and desiccation was repeated countless times, each round laying down another thin layer of sediment, until the rock built up into a vast block 300 m deep, obscuring the older, cratered landscape beneath. Although it is difficult to estimate how long this sea persisted, the thick sedimentary slab must have taken at least a quarter of a million years to accumulate.

Opportunity had discovered that not only did liquid water regularly cover hundreds of thousands of square kilometres across the Meridiani Planum but also that it had lasted for a relatively long time.

The evolution of life on Mars

What are the prospects of life evolving on Mars? The valley networks and Meridiani Sea suggest that primordial Mars was much like primordial Earth. Mars would also have received large deliveries of volatile substances and pre-formed organic molecules from meteoritic and cometary delivery during the bombardment. In fact, because Mars was closer to the snow line in the planetary disc it probably received more than Earth. This would make Mars, not Earth, the original water world of the inner solar system. Mars contains all the elements needed by terrestrial life. Some, such as iron, are more common there than in the Earth's crust, (the colour of the Red Planet essentially comes from plain old rust) while others, like iodine and potassium, are slightly rarer. The only vital element that today is present only in limited concentrations on Mars is nitrogen. Nitrogen in Martian surface minerals is biased towards its heavier isotope, which suggests that much of the original gas was lost with the rest of the atmosphere. However, nitrogen was probably not scarce on primordial Mars, so a lack of it would not have impeded prebiotic chemistry. Volcanism was rife on both young worlds, pumping out a thick carbon dioxide atmosphere to insulate the planet. The reducing nature of Mars' early atmosphere may have permitted Urey-Miller-type reactions, building up organic building blocks in surface pools. Hydrothermal systems could have existed in impact craters and surface pools above magma hotspots or in the sodden flanks of the great Martian volcanoes. Thus, at least in certain locations, Mars had the necessary basic conditions for life: liquid water, organic chemistry and a source of energy.

The Meridiani Sea was Mars' only long-running aqueous environment that we know about for sure. Although its water was salty, acidic and cold, terrestrial extremophiles can tolerate these conditions. The run-off from certain metal mines, such as in the Rio Tinto basin in Spain, is equally acidic and salty and contains cells living on minerals similar to those found at Meridiani. But few terrestrial organisms can put up with extreme salinity, acidity and cold all at the same time. The more difficult question is whether the desiccation and hyper-saline conditions created each time the sea dried up were survivable. We do not know how long the ground remained dry between floods but conditions in these intervals would have been very harsh in comparison to the wet times. Some Earth bacteria can endure short-term desiccation and spores encased within salt crystals for millions of years have been found to be viable. Therefore, it is plausible that Martian cells could have lain dormant until the floods returned and precious water soaked back into the soil.

What is less clear, however, is whether the intermittent wet spells were ever conducive to the evolution of life. Once life has become established, it can adapt and spread to more extreme environments, able actively to maintain its cellular interior at optimum conditions. However, the systems of prebiotic chemistry that produce complex organic molecules and lead to life may be far more delicate, unable to persist through the periods of increasing salinity and desiccation at Meridiani. None the less, for future missions, this dry seabed remains a very exciting place to visit. Sulphate deposits, like the sedimentary rocks found in Meridiani, are able to preserve organic molecules and the iron oxide deposits at Rio Tinto contain beautifully preserved, minute bacterial fossils. So if life did once exist in the shallow seas here, there is a good chance that we could find evidence of it, even billions of years later.

Some researchers believe they have already found proof of life on primordial Mars. The evidence comes from a single meteorite, a baseball-sized lump of Martian crust. Analysis of this rock has enabled much of its history to be pieced together. It was created deep underground, from cooling magma, very soon after the formation of Mars and as the original crust was solidifying. About four billion years ago, the rock was shocked and fractured by an enormous nearby impact and brought up to the surface. There are clear signs of water having trickled through the cracks a few billion years later, altering the chemistry and depositing carbonates and other minerals. Although this is evidence of aqueous alteration relatively late in Mars' history, the rock has mostly been untouched by the action of water, despite lying around near the surface for billions of years. This would be impossible if the environment was anything like the warm and wet terrestrial situation so the meteorite was probably only exposed to liquid water for a few hundred years. About fifteen million years ago, this chunk of rock was blasted off Mars by a nearby impact, which probably struck the ground at an oblique angle, ejecting the overlying rock upwards. Thirteen thousand years ago, the rock arrived on Earth, plummeting through the sky as a shooting star and falling on Antarctica, where it was quickly buried by ice and snow. Over time, glacial movements re-exposed the meteorite, its charred shell contrasting superbly with the snow-white background. This meteorite was collected by scientists in Antarctica's Allen Hills in 1984. It was the first that year and so was catalogued as ALH84001. A decade later, when the rock was identified as a piece of Mars, things started

Figure 11 Example of the bacteria-like microfossils found within the Martian meteorite ALH84001.

hotting up. The meteorite has been examined extensively; studies that led one group of researchers to assert that they had discovered Martian life within the rock. The team cited various pieces of evidence pointing towards biological action.

First, layered globules of minerals found within ALH84001 are similar to those produced by terrestrial bacteria as they metabolise. Second, a variety of organic molecules were found within it, including PAH (the carbon-ring molecules I described in Chapter 3, which are formed in interstellar dust clouds). The team argued that these PAH are very different from the kinds found inside the rock and so its PAH are not a contamination since the meteorite landed on Earth. PAH are not produced by terrestrial metabolism but are formed by the decay of organic matter. They are common in crude oil, for example, where they are formed by the breakdown of the chlorophyll of ancient photosynthetic organisms. So it is possible the PAH in ALH84001 represent the decayed remains of Martian bugs. Finally, microscope images of the meteorite's interior provided another suggestion of life. Elongated segmental forms (see Figure 11) are claimed to be the fossilised remains of Martian microbes. The problem is that these structures are tiny, some mere tens of nanometres in diameter, roughly the size of a virus and hundreds of times smaller than the smallest terrestrial bacterial fossils. It is uncertain whether all the basic machinery of life, such as a genetic polymer and enzymes, could even be stuffed into a package that small. However, the lower size limit on life is very poorly understood. The smallest accepted bacterium is some 300 nm across, and even smaller cells have been suggested in terrestrial rocks and implicated in causing heart disease.

The general criticism against the claims for life in ALH84001 is that none of the features are unambiguously biological – they could have been produced by abiotic processes. Combining different lines of ambiguous evidence doesn't make the case any stronger. The argument is very similar to the contentious issue of the first appearance of life on Earth. The jury is still out on

ALH84001 but much of the excitement surrounding it has dissipated. Many astrobiologists still believe in the possibility of
ancient, or even current, life on Mars. They just don't consider this
little meteorite to be convincing proof.

Environmental collapse

The valley networks and Meridiani sea date to roughly the same
time, very early in Martian history. Under current conditions
either would be impossible, for the water would either rapidly
freeze or sublime into the atmosphere. They indicate that once,
grossly different climatic conditions were present; a warmer wetter
Mars. The outflow channels were formed quite a bit later and
indicate very different climatic conditions again. Much of Mars'
water seems to have been locked in underground permafrost,
mobilised only by local geothermal activity that ruptured the ice
seal and liberated melt-water. Liquid water was only stable on the
surface when enormous volumes were spurted out in one go and
carved deep channels through the rock before disappearing again
into the ground or air. It is clear that around three and a half billion
years ago something very drastic befell the Martian environment.
Before then, the climate, while perhaps chilly, could none the less
support expanses of liquid surface water. Since then, conditions
have probably been much the same as the cold barren desert we see
today. Mars suffered a catastrophic environmental collapse while it
was still relatively young, an outcome believed to be directly linked
to the loss of its atmosphere.

Today, the Martian atmosphere is extremely thin but in primordial times, soon after the formation of both Mars and Earth, it is
thought that they were cloaked in similar atmospheres. Carbon
dioxide, methane and water vapour in the air would have provided
an appreciable greenhouse effect, insulating the young planets. The
combination of higher air pressure and raised temperature allowed

the long-term action of liquid water that we see signs of all over the Martian surface. To achieve the conditions to permit liquid water, Mars probably would have needed an atmosphere several times thicker than the current terrestrial blanket; it is further out than Earth and, in those early days, the Sun was also around twenty-five per cent dimmer. Much of this atmosphere would have been pumped out by the huge Martian volcanoes, driven by the heat stored within the planet's interior. Even during later times, when the outflow channels were created, each flood-triggering eruption might have released enough carbon dioxide for a slightly increased greenhouse effect, providing a brief spell of warmer conditions. But over time, the heat reserve within Mars dwindled and volcanic activity steadily diminished. The planet was dying inside and the eruptions replenishing the atmosphere with vital greenhouse gases became less and less frequent.

As the regeneration of the atmosphere decreased, the processes causing it to reduce continued unabated. Large impacts splashed away air, fast-moving gas molecules fled the gravitational grasp of the planet and the current of the solar wind gradually carried away the upper atmosphere. Although some of the gas would have reacted with the crust and been saved, the preponderance of the heavier nitrogen isotopes suggests that a great proportion was lost. With the erosion of the atmosphere, the surface pressure and temperature plummeted and the era of rivers and lakes came to an end.

It seems that, in many ways, this fate was inevitable. Mars is smaller than Earth, and has only about a tenth of its mass. This proved disastrous for its atmosphere in three ways: first, Mars has a much weaker gravitational hold over the gas molecules in its atmosphere and so they leak away more rapidly. Second, its smaller mass means that its store of internal heat runs down more quickly. Although there is some evidence of fairly recent volcanic activity at Olympus Mons, to all intents and purposes the Martian surface is geothermally dead and the atmosphere stopped being topped up long ago. Our own planet will suffer this same fate in perhaps another billion years. Third, the internal heat of Earth's churning iron core drives a powerful global magnetic field. This shields the Earth, to a certain extent, from cosmic radiation and deflects the solar wind, reducing the scavenging of the atmosphere. Mars is believed once to have possessed a strong magnetic field but for some reason it failed almost immediately, leaving the atmosphere unprotected.

What happened to Mars' water as the atmosphere bled away is less certain. The volume of ice visible in the polar ice caps corresponds to a few tens of metres of depth if the water were spread evenly over the whole surface but the minimum estimate of the amount needed to gouge out the drainage channels is several times that. Where has all the water gone? The hope is that much of it remains, underground, forming a hard permafrost shell many kilometres thick. Even if the evidence of a northern ocean does not hold out, perhaps because conditions at the time were too cold and each successive flood merely spilled across the frozen basin rather than topping up a liquid ocean, the vast volume of flood water would have soaked into the ground. The southern highlands, where most of the outflow channels originate, could also have much water still locked up in subsurface ice.

The top depth of this subsurface permafrost is generally thought to be a few hundred metres below the surface but to lie deeper around the equator, where the warmer temperatures have caused more of it to sublime. In some places, there are signs suggestive of liquid water having trickled across the surface in very recent times. Figure 10, for example, shows a series of gullies running down the face of a crater wall. Features like these demonstrate that not only is there water just beneath the Martian surface but that local geothermal activity melts pockets of it near the surface even today. The implications of this for astrobiology are enormous. Life is thought to need liquid water above all else and this is apparently present, at least for short periods, in the very top layer of the cold Martian ground.

Survival

If the Martian surface was once much more clement, what effects would the environmental collapse have had on emerging life? The collapse was probably fairly gradual and would have given surface organisms enough evolutionary time to adapt. Although endolithic communities (like those in the Antarctic dry valleys) may have clung on for a period as the environment around them collapsed, they must surely by now have succumbed to the extreme desiccation, cold and UV radiation. We may one day find their fossilised remains or discover decaying organic molecules as a calling card indicating they were once there. But so far, we have only checked in two locations on the entire face of Mars.

In 1976, the two Viking landing probes touched down on carefully-chosen sites on opposite sides of the Red Planet. Both carried a biology package, a miniaturised lab containing a set of experiments designed to test for the presence of life. The first involved scooping up a sample of soil and exposing it to carbon dioxide gas containing a radioactive isotope (^{14}C). After a short while, the soil was heated to drive off any volatile substances; encouragingly radioactive carbon was among them. This means that the carbon dioxide had become incorporated into something in the soil, much as terrestrial life fixes inorganic carbon by photo- or chemosynthesis. A second experiment fed another soil sample a radioactively-labelled nutritious broth of organic molecules. This time, the Viking landers found radioactive gas being given off by the soil, exactly as if life had been metabolising. A final test gave the soil a range of building blocks, such as sugars, amino acids and nucleotide bases and waited to see what gases were given off. Almost immediately, oxygen seeped out, as if dormant cells had been awoken by the warmth and moisture and begun photosynthesising. But the release diminished as more water was added, unlike what would happen with growing cells, and the same happened in the dark. This experiment, and the first one, also gave positive results even when the soil had been preheated to temperatures that would destroy any terrestrial cell. The initial euphoria among the Viking researchers rapidly turned to confused disappointment. The final blow came when some soil was heated beneath a chemical sniffer. Any complex organic molecules in the soil would decompose into simpler ones that would evaporate and could be identified. But the instrument couldn't detect even a whiff of anything organic, not even the breakdown products of dead cells. How could you have organisms without organics? Even without life, a certain level of organic substances was expected, due to the in-fall of comets and meteorites over Martian history, but the soil was totally lacking even this. What in the soil was performing the chemical reactions that on Earth would suggest life. And why is there absolutely no trace of organic substances on the surface?

The answer to both these puzzles probably lies in the same place. The thin Martian atmosphere, devoid of ozone or other filters, offers virtually no protection against the harsh UV radiation pouring from the Sun. This radiation soaks into the soil and creates highly reactive substances, including superoxides and other free radicals as well as powerful oxidants like hydrogen peroxide. These rapidly break and destroy any organic molecules in the soil; it is as if the Martian surface were laced with bleach. When the Viking experiments added water or nutrients, they too were split apart by the oxidants to release the oxygen and other gases detected. The Viking experiments are far from the final word on Martian life – all they have told us is that there is nothing at their two locations. The chemical analysis instrument also determined that organic substances could not be present at more than a few parts per billion. But this sensitivity would still miss cells, even if there were several million bacteria in every gramme of soil (roughly the density found in deep basalt communities on Earth); new packages, designed to be sent on future probes, are thousands of times more sensitive.

The harsh Martian conditions probably preclude any form of surface life. Solar UV, and the caustic chemicals it generates, is believed to have completely sterilised the topsoil. How far this oxidised region extends is not known but could be several metres deep, as the soil is continually mixed by the wind. The next generation of landers must burrow or drill below this lethal layer to have any hope of finding signs of life. As well as detecting the presence of any organic substances, astrobiologists are keen to know whether they might have been biologically produced. Future probes will look for two chief biosignatures: handedness and isotope ratios. As I described in Chapter 1, a property of many biological molecules is that they are always of the same handedness, so instruments will look for any bias. The second hallmark of metabolism is that it often favours the lighter isotope of elements, such as during the fixation of carbon or nitrogen; we'll return to this particular biosignature shortly.

Extremophile life can survive kilometres down in the Earth's crust, totally isolated from surface ecosystems. These SLiMEs live on geologically-produced gases dissolved in groundwater and are totally independent of the organic substances and oxygen produced by photosynthesic organisms above. Although the Martian surface is now devoid of life, could something be surviving deep underground? The raw materials for such chemoautotrophic life are basaltic rock immersed in liquid water containing dissolved carbon dioxide. The volcanic minerals in the rock are rich in highly reduced iron that reacts with the water to produce hydrogen, a fuel that the cells use both to fix carbon and provide energy from their redox reactions. Basaltic rock is thought to be common in the Martian crust but what about liquid water? The hard-frozen permafrost layer lying just beneath the surface is no environment for active cells but this reservoir could extend very deeply. The Martian crust is thought to be crumbly and porous to significant depths, due to the pounding it received during the heavy bombardment. Calculations show that this global sponge could hold an enormous amount of water, enough to create an ocean a kilometre deep if spread evenly over the surface but it is not known how much of this sponge is soaked. If the volume of water estimated to have formed the outflow channels and their sources in the southern highlands remains underground, then this permafrost shell is around 2 km thick at the equator and 6 km at the poles. At these greater depths, the interior warmth of the planet becomes significant and could have melted the bottom of the ice layer into slushy, partly-frozen, groundwater or even a briny liquid aquifer. Local pockets of liquid water could even be much closer to the surface, in the volcanic areas that are known to have been active in recent times. The Tharsis bulge was formed by a massive upwelling of magma and is such a heat source, relatively near to the surface. Reservoirs of salty water would provide similar niches to the pockets of water locked within Antarctic ice, inhabited by psychrohalophiles. Life on Mars may not form a planet-spanning ecosystem, as on Earth, but could be persisting in isolated refuges, where local geothermal heat melts the permafrost and leaches nutrients out of the rock. If Martian organisms are very rare, clumped in a few oases deep underground, how will we ever know where to look for them? The answer is that we might already have found the tell-tale signs of their existence.

Chemoautotrophs living in deep basalt aquifers react hydrogen and carbon dioxide to produce methane. On Earth, the release of this waste gas could never be used to hunt for deep ecosystems, because countless other biological processes also produce methane (including the hordes of bacteria dwelling in your gut). This is not true for Mars, so it is feasible that localised methane emissions could be detected in the thin air. One of the orbiting probes has indeed found trace levels of methane in the Martian atmosphere. This gas could have been released by small volcanic episodes or perhaps by the impact of a methane-containing comet but methane-producing bacteria are a third possibility. Methane is destroyed by solar UV and can last only a few centuries before being completely removed from the atmosphere, so whatever released this gas did so extremely recently and is probably still active. The instruments currently pointed at Mars are barely sensitive enough to have detected the methane in the first place and cannot accurately resolve where it is coming from but in the coming years, when we have put more landers on the surface, we should be able to detect the plumes as they rise into the atmosphere and locate their source. In principle, it ought also to be possible to demonstrate whether the methane has a biological origin. Biological processes using enzymes to facilitate chemical reactions leave a distinctive imprint: the compounds produced are generally biased towards the lightest isotope. For example, biological activity can be inferred in some sedimentary rocks on Earth because the carbon fixed by photosynthesis contains relatively more ^{12}C than it otherwise would. The same should be true of Martian methane if it were produced by biological fixation.

Aquifer water within the cracks and fissures beneath the permafrost shell is likely to be acidic, because of dissolved volcanic gases, and salty enough to remain liquid despite the cold. (It is important to recognise that although none of the extremophile organisms on Earth which tolerate such conditions release methane, this may simply be a reflection of the fact that cells in terrestrial environments have never been forced to evolve such a metabolic adaptation. There is no reason Martian cells could not adapt to this niche – no chemical grounds why energy couldn't be extracted using such a chemoautotrophic pathway under salty or acidic conditions.) In theory, SLiMEs could be lurking deep beneath the sterilised Martian ground in aquifers at the bottom of the permafrost shell. We may even be able to pin-point their locations under the surface from the tell-tale plumes of waste gas. The good news is that we might not have to drill through kilometres of solid rock to start looking for subterranean life. The ancient outflow channels were formed when geothermal hotspots melted through the permafrost shell to unleash cataclysmic deluges of water; there are signs that much smaller-scale events gushed water out on to the surface right up to the present day. For example, Figure 10 shows one such location, at the end of a set of fissures just north of the equator, littered with shattered slabs that closely resemble the pack-ice floes that drift Earth's polar seas. The theory is that underground permafrost was melted and flooded across the plain to form a lake about the size of the North Sea. The water froze but was prevented from subliming by a thin layer of dust and volcanic ash that fell over the area. The surprising fact is how recently this sea was formed – only about five million years ago. One hope is that subterranean cells were spat out on to the surface, along with the flood water, and were preserved as the sea froze. This presents a very exciting target to access Mars' subsurface habitat. Cells may even still be alive, frozen in suspended animation within the ice and could be retrieved in a sample return mission and revived in a lab on Earth. However, cells might struggle to remain viable after being dormant for so long under the relentless solar and cosmic radiation that beat down on to the unprotected Martian surface. There are terrestrial radiation-resistant bacteria that can tolerate such levels but only if they are active and can repair any damage as it happens. Cryo-preserved cells cannot; they could survive perhaps only ten million years before their DNA and proteins become so scrambled that they can never revive.

Mars has traditionally been the odds-on favourite for harbouring extra-terrestrial life, mostly due to its many similarities with Earth. But it is by no means the only possibile habitat in the solar system. With the advancing understanding of the past few decades, astrobiologists have become increasingly interested in other potential abodes. In the next chapter we'll tour these other planets and moons of Earth's family.

6

Elsewhere in the solar system

Astrobiologically, the potential of a planet or moon can be rated using three criteria: the existence of an energy source that can be exploited by life, the possibility of polymeric carbon chemistry and the presence of a liquid solvent. Life on Earth uses various energy sources, such as light or the chemical disequilibrium of reduced organic molecules or inorganic ions. Nutritious organic substances can also be created by electrical lightning discharges, inorganic fixation at hydrothermal vents, UV light, radioactive decay or ionising cosmic radiation and there may be enough sun-light to permit photosynthesis as far out as the Saturnian system. The occurrence of biologically-tappable energy sources is there-fore probably the least limiting of the three criteria. Organic mole-cules are rife within interstellar dust clouds and the formation of carbon polymers may be quite likely, given liquid water and a pool of reductants. And, discounting our earthly water chauvinism, we should also consider other solvents.

Keeping these criteria in mind, we can assess each body in the solar system for astrobiological potential. The central star, our Sun, was once believed by some to contain living beings. Proponents of this theory included William Herschel, the discoverer of Uranus. But we have since come to understand the blazing violence of its thermonuclear furnace; the Sun is clearly no dwelling place for any life we could imagine. Taking a tour out from the Sun, we first reach the inner rocky planets, Mercury, Venus, Earth and Mars. Of these, our world is the largest and only one with surface liquid water. Mercury is far too hot – baked dry by the overbearing Sun, devoid of water or organic substances – but Venus presents at least the hope of life.

Beyond Mars lies a ring of rock, the asteroid belt, believed to be rubble that was prevented from creating its own planet by the gravitational interference of Jupiter. Further out sit the gas giants Jupiter, Saturn, Uranus and Neptune, each formed from an ice-rock core that captured a thick atmosphere from the solar nebula. The ground level of the gas giants (if they have solid surfaces) is extremely hostile and presents impossible conditions for liquid solvents or organics. High amongst the clouds of Jupiter and Saturn, however, organic molecules are produced in Urey-Miller-like atmospheric reactions and some researchers have conjectured an aerial ecosystem. Heterotrophic, or possibly even photosynthetic, microbes could float on updraughts in the atmosphere, growing and reproducing quickly before they sink back down into the lethal heat and pressure of the interior. Other creatures, kept buoyant by great hydrogen sacs, might feed on this plankton, drifting among the clouds like airship-whales. However, this is little more than sci-fi speculation and the gas giants are generally discounted as likely abodes. The inner moons of the Jovian system are a stronger possibility; later, we'll explore the prospects for life on the ice world, Europa. If we relax the stipulation of liquid water and allow ourselves to consider other solvents and perhaps more exotic biochemistry, then Saturn's largest moon, Titan, is also a possible habitat. Beyond Neptune, where the solar nebula was thinner, a horde of small, Pluto-like, planetoids and vast numbers of icy comets lurks at the outskirts of the solar system. These can be instantly dismissed as potential habitats; they are extremely cold and lack conceivable liquid solvent or active chemistry.

So, of the 160 or so planets and moons within our solar system, astrobiologists have a shortlist of just four potential extraterrestrial habitats: Mars, Venus, Europa and Titan. Let's now explore the final three.

Venus

In early twentieth-century science fiction novels, Venus was often portrayed as a verdant swamp planet, hiding lush jungles beneath the thick mists of its atmosphere. This idyllic vision could hardly be further from the truth; the surface of Venus is utterly inhospitable. Venus is a textbook example of a world with a runaway greenhouse effect. Its atmosphere, composed almost entirely of carbon dioxide, is very thick, producing a crushing surface pressure ninety times greater than Earth's, creating a powerful greenhouse effect and an average ground temperature of almost 500 °C. This is hot enough to melt lead, which vaporises from the lowlands and recondenses as a metallic frost on the mountain peaks. The ground is a harsh surface of barren eroded rock, erupted from the interior during occasional intense bouts of volcanism. The little water in the atmosphere has dissolved these volcanic gases and occasionally rains down as concentrated sulphuric acid that re-evaporates in the burning heat before ever reaching the ground. At night (the Venusian night lasts almost sixty Earth days, due to Venus' slow rotation) the scorched ground glows dull red hot, as lightning flashes high in the eternally overcast sky.

Clearly no form of life is possible on this surface – no polymeric building blocks or liquid solvent could withstand these conditions. But that is not to say that astrobiologists have completely given up. Venus is Earth's twin; almost identically sized and would have received the same complement of volatile substances during the heavy bombardment. It is therefore very likely that, in their youth, the two planets were much alike, with similar atmospheres and oceans of water. Organic building blocks would have been delivered to Venus and it is possible that life got started. However, at some point, things started going very badly wrong. Closer to the brightening Sun, Venus warmed at an ever-increasing rate, until eventually its seas boiled dry. Water vapour rose high in the atmosphere, where it was split by solar UV into hydrogen and oxygen.

Hydrogen is too light a molecule for Venus' gravity to retain and it escaped into space, essentially stealing the oceans for ever. The oxygen reacted with surface rock and now there is barely any water left on Venus, not even in the atmosphere.

It is not known exactly when this catastrophe befell our twin but it could have occurred as late as two billion years into the planet's history; much longer than life took to emerge on the primordial Earth. Unfortunately, the face of Venus has long since been resurfaced by volcanism; there is no chance of finding evidence of ancient oceans or fossils of cells. Any trace of life on the ground has been completely smothered. Could Venusian life have adapted to survive this environmental catastrophe? There is one location on Venus that some astrobiologists think could still be inhabited. Conditions at a certain level within the cloud banks are not so different from Earth's surface. About 50 km above the ground, where the temperature has dropped to a balmy 40 °C and the atmospheric pressure is just seventy per cent of Earth sea level pressure, small amounts of liquid water are known to exist. It is not known just how strong the sulphuric acid concentration is but it is likely to be within the tolerance of terrestrial acidophiles. But could life persist in droplets of water suspended miles above the ground? Life thrives in practically all moist terrestrial environments but, although cells have been detected high in Earth's stratosphere, it is very dubious whether an ecosystem could be supported solely in clouds, without ever descending to the surface. If such a way of life were possible you might expect the terrestrial skies to have been colonised – why aren't Earth's clouds green with photosynthetic microbes?

If such an alien environment is tenable, it would have abundant energy from sunlight. If we view Venus not in the visible part of the light spectrum that we are most used to, but in the UV, something very curious stands out in the otherwise bland cloud layer. A complex and highly dynamic system of dark swirls contrasts with a lighter background, ranging from localised patterns to enormous planet-wide streaks. These dark regions show where UV light is being absorbed and account for almost half of the total solar energy soaked up by the planet. Could these be vast blooms of high-altitude Venusian microbes, using UV rather than visible light to photosynthesise? Probably not – but if biologists have learnt one lesson over the last fifty years it is not to underestimate the power of life to adapt to every possible niche.

Moons of fire and ice

Jupiter's four largest moons, Io, Europa, Ganymede and Callisto, can easily be seen with even the most inexpensive telescopes. They were discovered in 1610, when Galileo first turned his telescope to the heavens. This observation of satellites orbiting a planet other than the Earth helped to start the slow process of convincing the scholars of the Enlightenment that Earth was not in a unique situation at the centre of the universe. More recently, these four moons have forced a rethink of the original assumptions of astrobiology. The Galilean satellites formed beyond the snow line in the planetary disc, so they are rich in biologically-crucial volatile substances: Europa, Ganymede and Callisto contain large amounts of water. They are far beyond the habitable zone around the Sun and yet Europa and possibly also Ganymede, are strongly suspected to have liquid oceans. How can bodies so distant from the Sun's warmth sustain a liquid water environment?

Astrobiologists may have been a little unimaginative when first calculating the heat budgets of planets and moons. Beyond Mars, even a planet with a thick greenhouse atmosphere could not hold enough of the Sun's warmth to prevent water freezing but this is not the only heat supply available. The decay of radioactive elements in Earth's interior releases a lot of heat and the small Galilean moons have a third source: they orbit deep within Jupiter's powerful gravitational field, which exerts a strong tidal effect. Just as the

Moon causes a double bulge to travel around the Earth as it circles, seen most obviously in the oceans, Jupiter's moons are forced to flex as they orbit.

In the Earth–Moon system, the loss of energy from raising the tides resulted in the Moon's rotation slowing and it has become 'tidally locked': it rotates at the same rate it orbits the planet, so it always presents the same face to Earth and we never see the 'far side'. However, the gravitational interactions between the Galilean moons keep tugging at their orbits and prevent them becoming locked. The tides continue slightly to distort their innards. The moons are endlessly stretched and squashed, generating internal heat – just like a squash ball being repeatedly compressed during a game. The closer to Jupiter the moon orbits, the more tidal heating it experiences and the better the chance that some of its frozen water has melted into a liquid ocean. Io, the innermost moon, is so violently heated that it is in a state of perpetual global volcanism more active than the Earth's. Io's surface is coated an angry orange-yellow of erupted sulphur as this agitated moon is steadily turning itself inside out. The prospects of life amidst this thermal and chemical energy are slim; the heat would have long since destroyed any organic molecules and the volcanic plumes have ejected practically all the water into space. At the other end of the scale, Callisto orbits Jupiter almost three times further away than Io. Consequently, it receives much less tidal heating and there is no evidence of dynamic processes on its ancient icy face, heavily crater-scarred from the bombardment. Callisto is completely frozen and probably did not even properly differentiate during its formation; its interior is a jumble of rubble and ice. Europa and Ganymede, however, are thought to have a differentiated iron/silicate interior like Earth's, cloaked by a thick shell of water ice.

Between the two extremes represented by Io and Callisto lies an intermediate level of tidal heating. This would be sufficient to melt the bottom of the frozen shell and possibly to drive hydrothermal vents on the surface of the heated rocky core. This defines the

Figure 12 The orbits of Jupiter's largest four moons, Io, Europa, Ganymede and Callisto. There is a clear progression from the very fresh volcanic surface of Io to the ancient face of Callisto, and the the tidal habitable zone may lie somewhere in-between. For scale, Europa is roughly the size of our own Moon.

concept of an additional habitable zone; due not to solar warmth but tidal heating from orbiting a gas giant, as shown in Figure 12. Europa, second of the Galilean moons, is believed to be inhabitable for just this reason.

Europa

Europa is the smallest of the Galilean moons, roughly the size of our Moon. It is the smoothest body in the solar system, its surface a shiny shell of white to muddy-brown ice. There are relatively few

Figure 13 A region of chaotic terrain on Europa. The long parallel lines are believed to be cracks opened up by tidal flexing. The icy crust has subsequently partially melted and the great ice blocks reorientated before the surface refroze.

craters; this frozen crust is thought to be very young, no more than sixty million years old. The implication is that Europa is still an active world, regularly resurfacing itself, covering up old craters and the cracks produced by tidal stresses. There is good evidence that, a few tens of kilometres beneath the frozen surface, lies a layer of slushy ice or perhaps even a deep ocean of liquid water. Many visible features support the idea that occasionally this fluid gushes out on to the surface, before quickly refreezing in the −170 °C cold of deep space. Long parallel ridges seem to show cracks that have been forced open and re-closed by tidal stress, squeezing the welled-up water across the ground. Some regions even suggest hotspots melt completely through the ice shell to produce short-lived patches of open water. The photograph in Figure 13, for example, shows what looks like the intricately cracked surface fractured into large iceberg slabs that have drifted around before refreezing into place.

What are the prospects for life within this hypothetical global ocean, sealed beneath an icy shell? The best analogy on Earth for such a closed environment is provided by the group of a hundred or so lakes buried deep beneath the Antarctic ice sheet. Lake Vostok, by far the largest, is some 250 km by 50 km and up to 1 km deep. Although the Vostok research station has recorded the lowest temperatures anywhere on the surface of Earth (a numbing −90 °C) the lake is beneath 4 km of solid ice, where the temperatures are warm enough to sustain liquid water. The base of the ice sheet, well insulated by the thick layer of ice above, is melted by geothermal heat and the water settles into troughs in the underlying rock. Lake Vostok is particularly interesting because it is has been cut off from Earth's atmosphere for at least a million years. Any psychrophile life it supports will have been completely isolated from the rest of the biosphere for an appreciable length of time. Astrobiologists are keen to sample the lake to find what residents it might have but they need to be extremely careful not to contaminate its pristine waters. They have drilled down to within about 120 m of the top of the lake; a range of dormant bacteria has been found throughout this ice core. Safely penetrating actually into Lake Vostok will require more sophistication. One scheme involves a sterilised capsule with an internal heat source which will gradually melt its way down, allowing the ice to freeze behind it as it goes to keep the lake isolated. On breaking through, the capsule would release a 'hydrobot' to explore the frigid subterranean lake, analysing samples of the water for active cells. Just such a probe is being considered for a future mission to Europa. What are the chances that this isolated water supports life?

The potential Europan ocean would contain more water than there is on Earth and could provide a stable aqueous environment for the evolution of life. Two of three criteria I listed above, liquid water and organic chemistry, are probably present beneath Europa's surface. But what of an energy source to drive metabolism? Photosynthesis is impossible beneath such a thick layer of ice but chemoautotrophy is a distinct possibility. The Europan core may release reduced gases, such as hydrogen, that could be used in redox reactions to extract energy and fix inorganic carbon, just as in the basalt SLiMEs deep underground on Earth. Alternatively, tidal heating of the core might be sufficient to drive underwater volcanism and hydrothermal vents, releasing warm plumes of reduced inorganic ions into the sea. Some calculations put the heat flow within Europa at as much as a quarter of Earth's, comparable to Mars, and so subsurface volcanism is possible. Terrestrial black smokers along crustal spreading centres can support communities of chemoautotrophs and dependent heterotrophs. Such ecosystems, however, largely rely on dissolved oxygen from photosynthesis to release more energy from the redox reactions. This would obviously not be possible in the dark oceans within Europa, so alien black smoker communities would be perhaps ten thousand times less productive and support comparatively feeble ecosystems.

There are also serious doubts that the isolated ocean could provide enough oxidants to support life. For a cell to be able to extract energy from a redox reaction it must be able to pass the electrons released from the reduced fuel on to a more oxidised chemical. On Earth, even without photosynthesis, the atmosphere would ensure that a limited supply of oxidising power dissolves in the water, supporting chemoautotrophy. However, the Europan ocean is sealed; within a relatively restricted time-span any redox potential would reach equilibrium, destroying the energy gradient life depends on. How long this would take is hard to know but any long-lived ecosystem on Europa would need a steady supply of incoming oxidants – which the fracturing ice crust may provide.

Jupiter's powerful magnetic field traps the most intense radiation in the solar system. These energetic particles slam into the surface of Europa, ionising molecules trapped in the ice and giving them the energy to react together. Water is split to produce oxygen and hydrogen peroxide, two very oxidising chemicals that can be used to drive redox rections. The ice is also thought to be laced with carbon dioxide, which could be built up into simple organic molecules, such as formaldehyde. Terrestrial chemoautotrophs are known which react formaldehyde with oxygen; all Europan life would need is for this reservoir of vital molecules in the top ice to be transported into the ocean below. The signs of cracking and melting we see hint at just such a mechanism – the ice shell acting as a kind of solid atmosphere. This occasional mixing of surface ice with the water beneath would release floods of nutrients into the ocean, possibly triggering blooms of Europan microbes. In between such flurries of growth the cells may lie dormant, drifting through the cold ocean in suspended animation, waiting for the next flush of nutrients from above. In this way, the subsurface ocean could be periodically resupplied with oxidants and so the chemoautotrophic organisms effectively survive on space radiation. Whether this mechanism could provide a supply sufficient to support an ecosystem is highly contentious, though.

Directly spotting microscopic cells in the water would be difficult for a robotic probe, even if it could find hotspots of hydrothermal vents. A much better indication of biology might be to look for the tell-tale layered redox zones created by life's accelerated chemistry, as I explained in Chapter 1. We will have to wait many years before a probe exploring the Europan ocean begins sending back data but its findings could be as Earth-shattering as the discovery of Jupiter's moons were in the seventeenth century. How astounded would Galileo have been to know that, four hundred years after his first telescopic observations, this tiny pinprick of light was found to contain not only a deep ocean but an alien ecosystem?

From the Jovian system, our astrobiology tour takes us to the stunningly beautiful Saturn and its largest moon, Titan.

Titan

Saturn's moon Titan really deserves its name. It is the second biggest moon in the solar system, larger even than the planet Mercury. It is the only moon to have a dense atmosphere, five times thicker than Earth's and complete with smoggy haze and clouds. The air is mostly nitrogen, with a little methane; probably not far from the composition of Earth's primordial atmosphere. The geography of Titan is

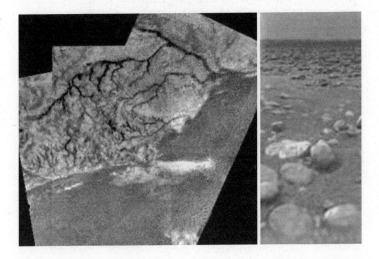

Figure 14 First images of a hidden world. Left: the surface of Titan seen by the Huygens probe as it descends through the haze, showing clear river valleys, the bright highlands, and darker bed of a dry lake. Right: pebbles of ice strewn around Huygens' landing site.

exquisitely diverse. As the *Huygens* lander slowly parachuted through the haze layer in 2005, we stole our first glimpse of the surface of this alien world, shown in Figure 14.

Titan appears disarmingly similar to Earth, with ranges of hills in the highlands, muddy-looking plains in the lowlands, fine networks of river valleys and a distinctly coast-like boundary between the highlands and plains. When *Huygens* reached the ground, its cameras showed a landscape strewn with pebbles, smoothed and polished by the action of flowing liquid. The lack of impact craters suggests an active surface like Earth's, with wind erosion, river processes and volcanism acting regularly to refresh the surface. The heat flow for volcanism would partly be provided by a small amount of tidal warming from Saturn but mostly from radioactive decay in Titan's core. This interior heat could melt the icy crust into a subsurface ocean of water, kept liquid with ammonia antifreeze and perhaps up to 200 km deep. The welling-up of this fluid through cracks in the crust produces *cryo-volcanism*, not erupting molten silicate rock like on Earth but a magma of a slushy water-ammonia mixture. There is also evidence of fault lines in the crust, suggesting the possibility of Titan-quakes and even plate tectonics.

In terms of active planetary processes, Titan is very much like the primordial Earth. The problem is that Titan is truly cold. It is almost ten times further out from the Sun and receives pitifully meagre warmth. The average ground temperature is some $-180\ ^\circ C$; there is simply no hope of liquid water under these conditions. The terrain visible on Titan is not built up of rock as we know it, the silicate rocks of the Earth's crust but of hard-frozen water – the hills and pebbles are made of ice. Rather than water, liquid methane is believed to be responsible for all the fluid action. Titan is like the Earth cooled down: atmospheric gases have condensed and flow as liquids across the surface and liquids like water have frozen solid to make up the rocks and mountains.

The hydrological cycle on Titan appears to be every bit as complex as Earth's, albeit that methane replaces water. Methane evaporates, saturating the air to form clouds that become laden and soon begin raining on to the highlands. The liquid methane runs downhill towards the plains, coursing along river valleys, picking up sediment, transporting rocks and smoothing them off, then pooling in shallow lakes before soaking into the ground. When *Huygens* touched down, it initially met a little resistance, before cracking through and sinking into softer ground. The interpretation is that the surface of Titan is like crème brulée; beneath a thin hard crust the ground is sodden with liquid methane. However, methane clouds are relatively rare in the hazy orange skies; the rains may occur more like a seasonal monsoon, coming every Titan year (about thirty Earth years). Despite this, warm lakes have been spotted near the North Pole.

The surface of Titan is so cold that the rate of chemical reactions is severely inhibited. Processes that are thought crucial to the pre-biotic chemistry of Earth, such as the polymerisation of hydrogen cyanide to produce amino acids, proceed on a timescale of tens of millions of years. So, although Titan's reducing atmosphere of nitrogen and methane may be favourable for Urey-Miller-like processes, organic reactions on the ground run agonisingly slowly. Titan is literally like the primordial Earth frozen in time. This is not to say that Titan is devoid of organic substances; it displays some quite interesting chemistry, although it is not yet known whether this builds anything as complex as amino acids or nucleotide bases.

Methane high in the Titan atmosphere is attacked by solar UV and energetic particles trapped in Saturn's radiation belt. Just as in the surface ice of Europa, the radiation ionises the molecules it strikes and gives them the energy to react. This drives the fixation of methane into a variety of more complex hydrocarbons. The larger molecules produce the thick orange haze that obscures Titan's surface and collect together into sooty particles. (An identical process to that which creates the photochemical smogs that smother large polluted cities such as Los Angeles or the ironically-named Buenos Aires.) On Titan, as these particles grow, they settle towards the ground as a gentle snowfall of hydrocarbons. The next methane downpour sweeps away these flakes, washing them downstream and flooding out into the wide plains where they collect as a thick sediment of organic sludge at the bottom of lakes. The lowlands are much darker than the hilly regions, probably because of a coating of hydrocarbon soot deposited by dried-up lakes. Dunes have also been spotted on Titan, which are probably made up of fine ice grains but could be made from granules of hydrocarbon polymers – plastic. It seems that Titan is not only physically active but its surface is also rich in biologically-relevant organic substances. If you could withstand the lethal cold of Titan, your nose would fill with the smells of an Earthly oil refinery: methane, ethane and other simple petrochemical organic molecules.

Could such a frigid and chemically bizarre world sustain life as we know it? There is certainly some geothermal heat, driving cryo-volcanism, as well as chemical energy available in drifts of hydrocarbons. There is a basic level of organic chemistry but we have no idea how far along the route of prebiotic chemistry this has progressed. Could such a cold environment allow the synthesis of long carbon polymers; replicating molecules for genetics or proteins for metabolic enzymes? There seem to be ample amounts of liquid methane but whether this could act as a biological solvent is not known. Without liquid water and its ability to donate oxygen atoms in reactions, could the simple organic compounds on Titan build up into sufficiently complex molecules? It has been proposed that nitrogen atoms might be satisfactorily swapped with oxygen in many biological molecules, donated perhaps by liquid ammonia. 'Amono-peptides' could possibly replace the protein polymers of amino acids. For the moment, we simply don't know enough about exotic biochemistries to postulate whether Titanic life really is plausible or not.

But perhaps Titan does not need unimaginably exotic biochemistry. Like Europa, Titan is believed to have subsurface liquid water. The colder temperatures on Titan mean this will probably be very slushy, even if mixed with ammonia, more analogous with the the viscous rock mantle beneath Earth's crust than the global ocean on Europa. Organic molecules created in the atmosphere and mixing with subsurface liquid water may be enough to support an internal biosphere. The water-ammonia mixture is calculated to be fairly alkaline at pH 11 and warmer than $-30\,°C$, with more favourable conditions around cryo-volcanic hotspots. The physical properties of this possible ocean are therefore within the tolerance of terrestrial extremophiles. For the first hundred million years of Titan's history, when it was still hot from its formation, even the surface would have been open ocean. This is the closest Titan has ever been to a true terrestrial environment and life may have started then. Enough sunlight was available for photosynthetic organisms but they would have become extinct as the ocean froze over and only chemoautotrophs could have survived.

What chemical energy might Titan chemoautotrophs feed on? One suggestion is the redox reaction between two compounds produced in the atmosphere: acetylene and hydrogen. Acetylene (the fuel used in welding torches) could accumulate in the sediment of hydrocarbon goo at the bottom of methane pools. Titan bugs could drive their metabolism from the energy released by reducing this acetylene with hydrogen, releasing methane as a waste product. If true, this astrobiological source could help explain one of the mysteries of Titan. Any methane in the atmosphere should be rapidly destroyed by solar UV, reacting to produce the sooty flakes of more complex hydrocarbons and all methane should have been completely removed within about ten million years. But instead, methane makes up five per cent of the Titan atmosphere – something must be releasing methane into the air. Cryo-volcanism or lakes are possible sources but chemoautotrophs could also be responsible. Some tentative evidence that supports the life theory is that the methane in the atmosphere is made up of slightly more of the lighter isotope of carbon than is predicted from some theories of Titan's formation. As I described in Chapter 4, this preference for lighter isotopes is a hallmark of the action of enzymes.

An alternative energy source proposed for Titanic life is rather more bizarre. The action of UV and particle radiation would produce free radicals, a class of extremely reactive chemicals. The recombination of free radicals provides a high yield of energy but terrestrial organisms find these chemicals impossible to control. In fact, they are the main cause of damage from aerobic metabolism and radiation. However, in the frigid conditions on Titan, these energetic reactions would proceed more slowly and could possibly be harnessed by life.

These last two chapters have looked in depth at the possibility of life on our planetary neighbour, Mars, and toured the other prime astrobiological locations in the solar system, Venus, Europa and Titan. But these are just the worlds on our doorstep – the galaxy is a very big place and presents a vast territory for astrobiology. So now, we'll turn our attention beyond our neighbourhood and towards the countless other suns spread across the heavens.

7

Extra-solar planets

Stars like our own

On a clear night, away from the light pollution of cities, the unaided human eye can see several thousand stars. The galaxy contains a few hundred billion individual suns. If merely a small fraction of these tiny pinpricks of light possesses a planetary system, the total number of planets in the galaxy is enormous. Our own local star, the Sun, is a fairly small, calm-mannered, middle-aged star but its remarkably stable nature might be exactly what has enabled life to emerge; astrobiologists are keen to identify similar stars in our neighbourhood.

Astronomers have identified two nearby stars that closely resemble the Sun. Tau Ceti is the nearest Sun-like star, twelve light-years away, clearly visible to the naked eye. It is twice as old as the Sun, at around ten billion years, but its planetary disc seems to be about twenty times dustier than our solar system's. This means that there are far more comets and asteroids; the Tau Ceti system could be plagued by a constant bombardment. A guardian gas giant might be able to protect the inner planets from killer impacts but the night skies of any Tau Cetian world would be permanently criss-crossed with comet tails and the streaks of meteorites hitting the atmosphere. Epsilon Eri is smaller, about eighty per cent the mass of the Sun. It is very young – it has yet to see its billionth birthday – so still deep in the heavy bombardment. If astronomers do find any planets within the habitable zone of this star life will not yet have had the chance to become established, due to being repeatedly snuffed out by mega-impacts.

Such Sun-like stars make up only a few per cent of the galaxy and astrobiologists are desperate to work out which other kinds might be suitable for supporting life within a family of planets. Astronomers classify stars by two basic qualities, their temperature and luminosity. The surface temperature of a star is reflected in the colour of light it emits. Hot stars are white or even blue, whereas cooler stars gleam orange or red. This spectrum is used to divide stars into different classes, each labelled with a letter. Running from hot to cool these are O, B, A, F, G, K and M (try the charming mnemonic 'Oh Be A Fine Girl, Kiss Me'). The luminosity of a star is linked to its size, with larger stars generally being brighter. The Sun is a relatively cool yellow colour and fairly small, making it a G-dwarf. Sirius, the brightest star in the sky, is an A-class giant and Betelgeuse, the crimson-coloured left shoulder of the constellation Orion, an M-class red supergiant.

I discussed the idea of the stellar habitable zone in Chapter 3. The HZ is the ring-shaped region around a star where liquid water is stable on a rocky planet's surface. Many different factors can affect it, such as the size of the planet and the composition of its atmosphere, but the basic concept is sound. The HZ can be calculated for any star in the galaxy and, as you would expect, it is wider and lies further out for hot stars and huddles much closer to cool stars. This relationship is shown in Figure 15, together with the corresponding stellar class of stable stars (that is, not including bloated giants) shown against the star mass and the positions of planets in our solar system. The HZ around hotter, more massive, stars is wider and one would expect these to represent more favourable astrobiological targets. The major problem with hotter stars is that they guzzle hungrily through their nuclear fuel. The most rapidly-paced stars last only a few million years – far too short a period for life to emerge on rocky inner planets. Large stars also violently explode as supernovae, their shock wave shattering any inner planets and char-grilling the moons of gas giants. Thus, the hottest stars, O and B-class, are not generally considered capable of

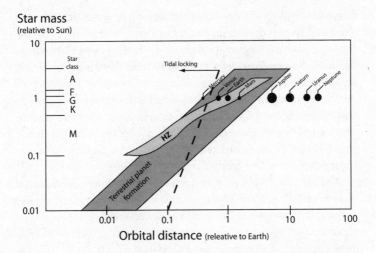

Figure 15 Diagram showing the extent of the Habitable Zone for different classes of stars. The positions of planets in our solar system are shown, with only Earth squarely within the HZ. Planets too close to their star become tidally locked after a few billion years, and for M-class dwarf stars the entire HZ is within this boundary.

supporting life in any of the potential niches previously described in our solar system.

Red dwarves

Stars less massive than the Sun present a much more intriguing possibility for astrobiology. The smallest, M-class dwarves, are much cooler than the Sun and glow with a dim red light. They are so feeble that even the closest and brightest in our skies, Proxima Centauri, is hundreds of times too faint to see by eye. But such stars are much more numerous than heavier Sun-like ones, so if they could support life on rocky planets they represent an enormous number of possible oases in the galaxy. They are so thrifty with their fuel resources, sipping through hydrogen like miserly stellar Scrooges, that no M-class dwarf has ever died. They can last *trillions* of years; the very first ones to form in the early universe are still going strong today. This slow and steady routine gives life plenty of time to evolve and so it might seem that red dwarves are even better than Sun-like stars for the prospects of astrobiology. But, unfortunately, they're not ideal nurseries – they present quite a few problems and astrobiologists are unsure for the moment whether or not these are show-stoppers.

The HZ of an M-class star is about five times narrower than that around the Sun. This is quite a small target for rocky planets to form and maintain a stable orbit within over billions of years but there is an even more serious concern with the red dwarf HZ. Figure 15 shows two other crucial aspects of a planetary system. The shaded stripe represents the region around a star where rocky terrestrial planets can form. Planets do not form less than than a certain distance from the star because the dusty material in the solar nebula cannot aggregate. Further out, beyond the snow line of volatile condensation, planetary embryos form from ice as well as rock, their cores rapidly growing massive enough to sweep up great volumes of gas and create giant planets with surfaces unsuitable for life. The dotted line marks the limit of tidal locking. If a planet orbits too closely, it gradually loses its rotational momentum to tides raised by the star and within a few billion years becomes tidally-locked, always presenting the same face inwards. Figure 15 shows that Mercury is within this threshold and so has a locked rotation. Venus is also close and has a greatly retarded rotation, with a single Venusian day lasting 117 Earth days. For M-class stars the entire HZ is inside this limit – any habitable planets become tidally-locked relatively quickly. For the very coolest red dwarves, the majority of the habitable zone falls outside the region of terrestrial planet formation, making life all but impossible.

The consequences of tidal locking on a planet orbiting a red dwarf are profound. One side of the world will always face its star and be eternally scorched, while the far side never sees the light of day. Much of both sides is likely to be utterly inhospitable: the facing side baked dry by the perpetual sun looming blood red in the sky and the far side so bitterly cold that the air freezes and snows down to the barren ground. In between these two extremes there could exist a region, circling the world, where liquid water and perhaps even life itself would be possible, a twilit zone where the sun hangs eternally on the horizon. To have any real chance of life, a planet fixed by its sun would need a thick atmosphere to redistribute the heat it receives. This would generate a truly fearsome wind, as planet-wide convection currents rebalance the temperature gradient, with hot air from the sun-side expanding and rushing around to the dark-side. However, it may be difficult for a planet orbiting so close to its star to retain a thick atmosphere. Mars is thought to have lost most of its original atmosphere, some blown away by the solar wind. Red dwarf planets may be especially susceptible and might need to be particularly large, to keep a firm gravitational grip on their atmosphere, and have a powerful magnetic field to deflect the solar wind.

Furthermore, M-class dwarves have quite an agitated temper. They are much more active in solar flares and sunspots than the Sun and their brightness can rapidly vary by as much as ten per cent. By comparison, our Sun has very gently increased in brightness, by about twenty-five per cent, over four billion years. The effects of a variable light output are difficult to model but are almost certainly very destabilising to the planetary feedback systems I discussed in Chapter 3. The climate of a red dwarf world could swing between greenhouses and glaciations. The solar flares may be equally problematic to surface life, as UV radiation levels would spike enormously every time a solar flare licked up towards the planet. A thicker ozone shield than the Earth possesses, or other UV filters such as the photochemical haze on Titan, could adequately protect the planet. Another feedback loop may naturally serve to control UV levels: ozone is produced by UV light splitting apart oxygen molecules, which can be released either by photosynthesis or through the action of UV on water vapour in the air. So, the higher the UV levels hitting the planet's atmosphere, the more ozone is produced to absorb it. Even without a complete ozone shield, life around a red dwarf could take refuge underground, within fractured rocks or produce large amounts of biological UV screens, as do terrestrial lichens.

Red dwarves become much more tranquil and cordial as they age; flares and sunspots are only likely to be a major hazard for the first billion years or so. Earth-sized worlds would still be volcanically-active enough, after this tempestuous teething-period, to regenerate a thick insulating atmosphere and have a further few hundred billion years leisurely to evolve life. However, a tidally-locked world orbiting a red dwarf star is such an unknown situation that astrobiologists are divided as to whether they represent worthwhile targets or not. Most of the forthcoming hunts for potentially life-bearing planets are focussed on more Sun-like stars.

Planet hunters

Although the distribution of nearby Sun-like stars has been known for some time, only fairly recently have we been able to address the question of how commonly stars also posses a system of planets. A number of young stars have been spotted with dusty discs but the first detection of actual planets orbiting another star came in 1991, around a *pulsar*. Pulsars are a type of star that emit a tight beam of radio waves that sweeps through space as the star rotates, much like the beam of a lighthouse. If Earth lies in the path of this beam, astronomers can pick up a rapidly beating signal, like a celestial metronome (pulsars are some of the most regular objects in the universe). The gravitational tug of an orbiting planet slightly displaces the position of the pulsar, so that its distance from Earth alters a little: even though the pulsar is beating at exactly the same rate, the precise timing of the radio pulses varies as the star wobbles. The effect is tiny but because the pulsar has such a regular and rapid tick (every six-thousandth of a second or so), it can still be detected. This allowed astronomers to infer that two planets of several Earth-masses, and one much smaller, were in orbit around it.

But although pulsars are ideal for detecting planets, they are one of the least likely kinds of star to find life around. Pulsars are formed from the collapse of a stellar core after it has exploded as a supernova. No terrestrial planet could survive this, so the bodies orbiting the pulsar must have re-formed from the shattered débris left over from the explosion. These planets will be charred husks of rock, completely sterile. What astrobiologists need is a suite of other techniques to detect planets around more Sun-like stars.

As of the time of writing, 168 planets have been discovered orbiting alien suns and more are being discovered at an ever-quicker rate. The closest orbits a star just over ten light years away from us, the furthest one 17,000 light years away. Eighteen stars possess families of several planets. There are four methods that can be used to search for extra-solar planets, as shown in Figure 16, although the last has yet to make its first discovery.

The most successful by far is the *radial velocity technique*, which has been used to find around ninety per cent of the known extra-solar planets. As an object approaches an observer, the light it emits is shifted to higher frequencies (is *blue-shifted*) and as it recedes the light is shifted to lower frequencies (the *red-shift*). This is the *Doppler effect* and is commonly experienced as the characteristic rise and fall in pitch of an ambulance's siren as it overtakes you. The Doppler shift of the light emitted by a star tells us about its veloc-ity relative to Earth. The radial velocity technique looks for a cyclic change in velocity as the star is pulled slightly by an orbiting planet. The magnitude and timing of this effect reveal the mass and orbit of the planet causing it. But only a lower boundary can be put on

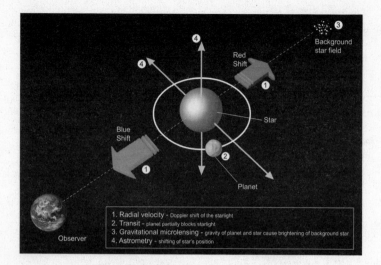

Figure 16 Four methods for detecting extrasolar planets: radial velocity, transit, gravitational microlensing and astrometry.

the planet's mass, because it is not known how the orbital plane is angled, relative to Earth. For example, if the planet's orbit happens to be aligned along the line-of-sight from Earth, so that we view the system edge-on, even a planet of minimum mass would be enough to tug the star towards us with the observed velocity. If the planetary system were oriented much more face-on, a more massive planet would be required to produce the measured shift.

The *transit method* involves detecting the periodic dip in the apparent brightness of a star as a planet passes in front and eclipses it. This method obviously only works if the plane of the planet's orbit happens to line up with our line-of-sight, so that we view the system edge-on.

A third scheme, *gravitational microlensing*, exploits an interesting aspect of Einstein's theory of relativity. This relies on the fact that the gravity of a body can focus the light of an object behind it. If astronomers point their telescopes at a region of sky with lots of stars, such as towards the centre of the galaxy, every now and then one of them might slowly brighten and dim as another, unseen, star passes exactly between Earth and the distant star. Any planets orbiting the middle star, the lens, cause their own slight brightening, a blip on the light curve.

The fourth method, *astrometry*, requires extremely precise measurements of a star's position in the sky. As a planet orbits, its gravitational tug causes the star to shift slightly around their centre of mass. Such a wobble in the star's position theoretically allows the mass and orbit of any planets to be calculated but our instruments aren't yet quite accurate enough for this technique to work practically.

Hot Jupiters

Figure 17 shows a summary of just a few of the extra-solar planets, or *exoplanets*, discovered to date, compared to our solar system. Most of these planetary systems aren't much like ours. Jupiter, the Sun's largest planet, orbits at 5.2 times the distance Earth orbits (and doesn't even fit on the diagram) but many of these other planetary systems have planets larger than Jupiter in their inner system. For example, one of the first extra-solar planets to be discovered is over four times more massive than Jupiter and orbits its star, Tau Bootis, at a tenth of the distance that Mercury does the Sun. This giant orbits so closely that its year lasts just three Earth days and its atmospheric temperature is a blistering 1,000 °C – hot enough to melt the change in your pocket. Such extra-solar planets have come to be called 'Hot Jupiters' and their very existence is puzzling. The stand-ard theory of gas giant formation is that the accretion of icy-rock small planetary fragments in the outer system allowed these more massive cores to gather heavy atmospheres of gas from the planetary disc. If this model is correct – although it is,

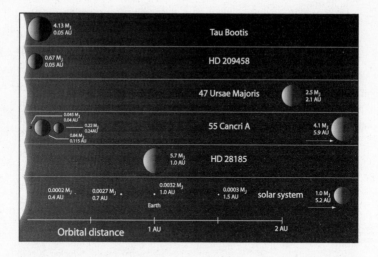

Figure 17 Schematic of some of the alien planetary systems discovered so far shown compared to our own solar system. Planet masses and orbital distance are given relative to Jupiter's mass and Earth's orbit (1 AU).

admittedly, based only on the example of our solar system – it means that gas giants simply cannot form inside the snow line, so near to their parent star. Hot Jupiters ought to be impossible but they account for many of the extra-solar planets discovered to date.

Perhaps these Hot Jupiters didn't form so close to their star but have migrated from their birthplaces in the outer system into much tighter orbits. In theory, a gas giant could steadily lose orbital energy as its gravity interacted with the remains of the accretion disc and so steadily spiral in towards the star. This orbital decay would probably be disastrous for any inner planets then forming, including those fortuitously within the star's habitable zone. The gravitational disruption of the marauding giant would seriously upset their orbits, possibly dragging them into collision, ejecting them out of the system to wander endlessly through the cold void of interstellar space or plunging them headlong into the thermonuclear inferno of the star. There is evidence, based on the presence of heavier elements in a star's atmosphere, of planets being swallowed by their own sun. Once the planetary system has been cleared of the dust, the migration stops (so we are safe from a Jovian rampage). There are indications, however, that Jupiter did migrate from its birthplace, by perhaps ten per cent, during the early period of the solar system.

Even if a gas giant were within the habitable zone of the star its surface would be completely inhospitable. None of the extra-solar planets yet detected could possibly be abodes for life and, what's more, the hot Jupiters have probably dashed the hopes of any terrestrial planets within the HZ. But any moons orbiting these gas giants, worlds like Europa or Titan transported closer to the solar warmth, represent a very intriguing possibility. For these displaced moons to be inhabitable, they would need to satisfy many of the same requirements as a planet. To give life a fighting chance, the giant would first, need to remain stable in the habitable zone for billions of years and second, not have a very eccentric orbit, to minimise climatic swings. Energy sources on such satellites are unlikely to be limiting factors, as they could be heated from inside by tidal forces from their gas giant and from above by the star, making autotrophy both around hydrothermal vents and through photosynthesis plentiful. And because the moon formed beyond the snow line before being dragged inwards as its giant migrated it will be loaded with volatile substances, both organic molecules and the ingredients to form atmosphere and ocean. However, there could be too much of a good thing: the Jovian moons Ganymede and Callisto are composed of at least fifty per cent water ice; if Jupiter had migrated all the way into the Sun's HZ, this would have melted into a global ocean over a hundred kilometres deep. The crushing pressure at the bottom of such an ocean might preclude life around any hydrothermal vents and although photosynthetic bacteria might happily drift through the surface waters, it is unclear whether prebiotic chemistry could ever produce life under these circumstances. Earth's climatic thermostat, the carbonate-silicate cycle, would also not function on a moon with no land surface. These moons would also become tidally locked to their planet and so rotate very slowly relative to the star, presenting similar problems to red dwarf planets. A thick atmosphere could redistribute the uneven heat but to hold on to this a moon would need to be at least three times bigger than Ganymede, the largest moon in our solar system. None of these may be critical problems and life on moons orbiting gas giants remains a fascinating possibility. Indeed, astronomers think they may already have detected just such a suitable gas giant.

If you live in the Northern Hemisphere take a step into your back garden on a clear winter's night (a summer night in the Southern Hemisphere) and gaze towards the famous constellation of Orion. Snaking away from the outline of the hunter runs Eridanus, the sacred river; within the fold of the first meander of this celestial stream lies a star with the unassuming name HD28185. Unfortunately HD28185 is too dim to see with the naked eye but its position is shown in Figure 18. This star is a yellow G-dwarf, slightly cooler than our Sun; in 2001, astronomers using the radial velocity technique discovered it has a planet. This world is some five times more massive than Jupiter and so is not thought able to cradle life but it has a circular orbit neatly within the habitable zone of its star (its year lasts only twenty days longer than Earth's), so liquid water and life on orbiting moons are a possibility. Although no such moons have yet been discovered, astronomers consider this system to be one of the best candidates for alien life.

Before the discovery of migrated gas giants, astrobiologists had not even considered the possibility of large moons within the habitable zone. Another eye-opener countered the assumption that planets could only ever form around solitary stars. Our Sun is a singleton, formed at the same time as other stars within the

Figure 18 The constellation Eridanus lies next to the more familiar Orion. Indicated is the location of HD28185, the Sun-like star whose planet orbits neatly within the habitable zone. If this gas giant possesses any moons they could well bear life.

interstellar cloud but not gravitationally bound to anything else. However, two-thirds of the stars in the galaxy have at least one companion; the closest star to the Sun, Proxima Centauri, is part of a triple system, orbiting Alpha Centauri A and B, a pair of stars not unlike the Sun. The Sun may be somewhat special in being alone and it was thought this might be a crucial factor in its habitability. Astronomers had calculated which sorts of planetary orbits within a binary star system are stable. For example, a planet could closely orbit just one of the pair or be set on a wide circle around both tightly-bound stars but a figure-of-eight route around both stars is unstable and such a planet would ultimately be flung from the system. However, it was thought impossible for planets to ever form around two stars, as the accretion disc would be swept away by their gravitational interplay. The discovery, in 2005, of a Hot Jupiter within a triple star system came as a great surprise, serenely circling its Sun-like parent while the other pair orbits on an out-side path. It is not clear how this unlikely planet could have formed but it must experience some breath-taking sunsets, as red, orange and white suns continually arc through its skies.

The most extensive extra-solar planetary family known is that of the star 55 Cancri A. Not only does this system boast one of the smallest planets yet discovered, a Neptune-sized body orbiting close to its star, but another three gas giants further out in the system. If a fifth planet is discovered, we will know as much about this completely alien planetary system as we did about our own solar system for the first 4,000 years or so of astronomy, while Saturn was the last 'wandering star'.

But the most striking feature of the extra-solar systems discov-ered so far is how different they are to ours. This is probably not due to a lucky quirk of our situation but is an artefact of the way we are currently looking for planets. The techniques described above indirectly detect the presence of a planet by its effect on the light, position or velocity of its parent star. It's no surprise that the present sensitivity of our equipment has only picked up very large planets orbiting very closely to their star. When Tau Bootis was dis-covered in 1996, using the radial velocity technique, the corre-sponding shift was 470 metres per second. Leading research groups are currently achieving a precision of 3 metres per second – the speed of a gentle run – which equates to the effect that Saturn has on the Sun. This illustrates another important facet of extra-solar planetary detection: Saturn has an orbital period of thirty years and it takes astronomers at least one cycle to be sure of the mass and orbit of the planet they have discovered. Astronomers have simply not yet been observing for long enough to detect anything other

than giants huddling close to their star. However, the next generation of telescopes, are designed specifically to hunt out terrestrial
planets within the habitable zone of their stars.

Next Generation

By virtue of their small size, terrestrial planets are extremely difficult to spot by their effect on the parent star; the search for them is
going to require some of the most sensitive telescopes ever built.
The turbulence of the Earth's atmosphere would largely swamp
any signal from an orbiting habitable world, so most of the next
generation of planet-hunting telescopes are to be positioned in
space. *Kepler*, scheduled for launch in around 2020, is designed to
use the transit method, spotting small rocky planets as they pass
across their sun. It will observe 100,000 stars, taking brightness
measurements every quarter of an hour for three years. Although
this technique can only detect planets orientated on line-of-sight,
it is still hoped that this sample will pick out around fifty Earth-like
planets. Another project, the *Space Interferometry Mission* (SIM), will
attempt to detect terrestrial worlds by astrometry, as the planet
causes the star's position to wobble ever so slightly. This task will
require measuring the position of the star extremely precisely, as
the change in angle to the star is equivalent to the diameter of a cell
on Earth viewed from the Moon. SIM will achieve this astounding
feat using *interferometry*.

SIM will collect light from the target star using two mirrors nine
metres apart. This allows the telescope to emulate having a single
enormous mirror the diameter of that baseline distance; giving SIM
the same resolving power but still obviously collecting much less
light. The beams from the two mirrors are combined so that the
light waves interfere with each other, producing a characteristic
strip of light and dark spots, or *interference pattern*, where the peaks
and troughs of the waves combine or cancel. This pattern is

extremely sensitive to the distance the light has travelled before it enters the mirrors, so as the star wobbles with an orbiting planet, the interference pattern it creates measurably shifts. SIM will use this ingenious technique both to check over two thousand stars of a wide variety of classes at low precision and focussing more closely on seventy-five of the nearest Sun-like stars. This will provide us with information on the diversity of planetary systems and hopefully help to find a few terrestrial worlds within the habitable zone.

This technique of emulating colossal telescopes by arrays of smaller collectors can be extended as far as you like. There are several plans for ground-based arrays of telescopes with baselines of up to 200 m. But if you want a truly enormous array, unhindered by the shimmering and absorption of the atmosphere, you need to be in space. Both NASA and ESA have plans to launch space-based interferometry arrays, respectively the *Terrestrial Planet Finder* (TPF) and *Darwin*. These missions will go far beyond detecting a planet indirectly and will actually gather the light from an alien earth. The two designs differ slightly in details but both involve fleets of independent spacecraft, each with their own mirror, positioned along a baseline of a few hundred metres, linked to a hub-station where the light beams will be combined. The technological difficulties are astounding: each telescope craft must point in precisely the same direction as well as keeping its position within the formation to an accuracy of tens of nanometres.

Attempting to see an extra-solar planet directly is difficult for two reasons: not only is it very close to the star but its reflected light is about a billion times dimmer and is utterly swamped. It's like trying to spot a glow worm crawling around the rim of a searchlight. The trick is to look not in the visible part of the spectrum but in the infra-red (IR) 'thermal' light. The hotter an object, the higher the frequency of light it emits. The Sun gives off mainly visible light and a little UV, whereas the Earth emits infra-red light from the warmth it has absorbed. By looking for terrestrial planets using infra-red, astronomers are able to decrease this contrast to 'only' a million times different. But as well as allowing these arrays to emulate an enormous single telescope, interferometry has another trick up its sleeve. The beams from the separate mirrors can be recombined in such a way that the light in the centre of the visual field is made to cancel itself out. Light rays coming from slightly off-centre, however, such as from a planet orbiting nearby, are not affected. This allows the telescope effectively to turn down the brightness of the star, so that the planet stands out.

Light of a distant earth

TPF and *Darwin* will detect extra-solar planets by isolating their light from the glare of the parent star and an enormous amount of information can be gleaned about a world by analysing this dim glow. Regular variations in the intensity and frequencies of the light emitted reveal the day length; the difference between expanses of water and land shows up in infra-red and so as the planet rotates coarse features of the geography can be seen, revealing the proportion of continents and oceans; desert sand, ice and cloud cover may even be distinguished and so some idea of the regional climate can be gauged; and as the planet cycles in its orbit, seasonal variations in cloud or ice cover might be detected. However, for astrobiologists, the most exciting possibility by far will be actually to read the chemical composition of the planet's atmosphere.

The basic technique of reading chemical composition from transmitted light has been around for well over a century. Light is passed through a spectroscope – a device which splits light according to its wavelength, just as droplets of water produce a rainbow. Light emitted or absorbed by different substances produces characteristic patterns in the spectroscope. For example, atomic gases absorb light in the visible region of the spectrum. In 1868, the light of a solar eclipse was passed through a spectroscope and showed the absorption pattern of a substance never before seen on Earth. This new element was named helium, after the Greek, *helios*, Sun.

Many molecules absorb infra-red light, where the energy levels correspond to different modes of vibration in the bonds between atoms. Each type of bond absorbs a particular wavelength of IR, so a graph of the infra-red spectrum will show dips where light has been absorbed. What molecules are present can be worked out from the appearance of different types of bonds so the absorption spectrum is the molecules' 'fingerprint'. By spectroscopically analysing the infra-red light collected from an extra-solar planet, astronomers can identify the molecules present in its atmosphere. Visible light spectroscopy has already been performed on starlight passing through the atmosphere of an extra-solar planet, HD209458b, discovered by the transit method in 1999. It is one of the most extreme Hot Jupiters known, orbiting so closely to its Sun-like star that its atmosphere is being blasted away into space, flowing behind like a comet's tail. Astronomers have estimated that it is losing at least ten thousand tonnes of hydrogen every second. This inflated puff of atmosphere allowed the Hubble Space Telescope to perform spectroscopy as HD209458b passed in front of its star, revealing the presence of sodium by its absorption fingerprint in the visible part of the spectrum. This result is not particularly informative in itself, as sodium is thought to constitute merely a tiny amount of the atmosphere and was only detected due to its obvious spectroscopy signal but it is very significant as a proof-of-principle. Once the large space telescopes are operational we have every hope of analysing the composition of alien atmospheres. But what exactly would it be about the composition of a planet's air that would represent a sign of life? How can we tell that a world light years away is not just habitable but is actually inhabited?

The original idea dates to the 1960s, when NASA was trying to work out how life might be detected on other planets. One proposal involved using spectroscopy to analyse the chemical composition of the atmosphere, an experiment that can be performed using a telescope without even needing to send equipment to the planet. The point is that the atmosphere of an inhabited planet is expected to have been pushed far from chemical equilibrium. An ecosystem interacts with the rock, water and air of the planet; the very action of life leaves an imprint that betrays its existence. This idea has been extended over the years and come to be known as the *Gaia Hypothesis*, after the Greek goddess of the Earth.

Gaia

Life develops into great interconnected webs of species: ecosystems. Supporting these living networks are the primary producers, extracting energy from the environment to run their metabolisms and reproduction. Different mechanisms of primary production include photosynthesis in surface ecologies and chemosynthesis in hydrothermal vents or deep basalts. Other organisms, heterotrophs, live off the creative efforts of the producers as herbivores or predators; sometimes the two form a mutually symbiotic relationship, like the tube worms and their chemoautotroph companions around black smokers. The most productive ecosystems are dense hierarchies of animals eating plants, animals eating animals, plants parasitising on other plants and plants and animals exploiting the return of nutrients to the air and soil by decomposers such as fungi and bacteria. Similar cycles of nutrient and energy flow exist in the open oceans and to a certain extent in the isolated ecosystems huddled round hydrothermal vents. To paraphrase, no organism is an island, entire of itself but a piece of the biosphere, a part of the main.

The biosphere is just as dependent on the abiotic components of the Earth. The raw materials of life, carbon, nitrogen and oxygen, are endlessly recycled through the rocks, oceans and air of the planet. In particular, the atmosphere acts as a reservoir of resources, such as oxygen for aerobic cells and as a repository for waste products, such as carbon dioxide from respiration. The essence of the

Gaia theory is that not only do the biotic and abiotic spheres interact but that they do so in a way that each apparently regulates and balances the other. The entire system is stablised: some proponents even go so far as to describe the Earth as a single 'superorganism' that operates to maintain its own stability. There is no teleological argument here: the Earth does not follow any design; the biotic and abiotic spheres do not 'intend' or 'strive' to balance each other. The Earth is an incredibly complex, interconnected system within which certain feedback loops have spontaneously arisen to control and limit change away from equilibrium.

I have already discussed one of these feedback loops in detail. The carbonate-silicate cycle acts to stabilise the Earth's climate, balancing the opposing forces of rock erosion and carbon dioxide levels. Life has become incorporated into this vital control system. Many organisms in the ocean act to raise the rate of carbon storage, using carbonates to construct their shells, which is locked into limestone deposits when they die. Soil bacteria accelerate the crumbling of rocks on land, an effect that is increased by warmer temperatures and the film of photosynthetic organisms covering the planet fix carbon dioxide from the atmosphere. Photosynthesis has increased the redox potential of the whole Earth, pumping oxygen into the atmosphere and depositing highly reduced organic matter into ground and sea sediments. Biological processes can have effects on a truly global scale and can transform the conditions and composition of an entire planet. But how can the action of life be distinguished from widespread abiotic chemistry?

The signature of life

There is something very special about many of the processes associated with life. The inevitable trend in any system is of decay towards increasing disorder. Living things, as distinct from abiotic arrangements of atoms, are staggeringly complex and ordered. Such structures cannot arise spontaneously but must be created through carefully controlled reactions. Life must extract energy from its environment to drive certain reactions, creating complex molecules from simple precursors. Plants, for example, derive energy from sunlight to convert carbon dioxide and water into large carbohydrate molecules like starch, releasing oxygen in the process. Oxygen is very reactive and quickly forms oxidised compounds. So, the production of oxygen by photosynthesis is indicative of these chemical reactions that are crucial for life. An oxygen–rich atmosphere is said to be out of chemical equilibrium, as the oxygen must be being produced as quickly as it is removed. One of the necessary side-effects of life is that it modifies the chemistry of its surroundings in ways that might not otherwise have happened.

An oxygen-rich atmosphere is not in itself a good indicator of life. Several inorganic processes are also known to produce this gas. For example, the runaway greenhouse effect that befell Venus would have boiled large amounts of water into the upper atmosphere, which would have been split by solar UV into its two component gases with the light hydrogen quickly leaking out of the atmosphere leaving relatively high levels of oxygen. However, high levels of oxygen, when combined with the presence of reducing gases, is considered to be a good biosign. To keep pace with their destruction by oxidation, these reducing gases must be constantly created. Methane, for example, is quickly oxidised to carbon dioxide and water; to maintain the levels in Earth's atmosphere far from chemical equilibrium, over a billion tonnes of it must be produced every year.

Oxygen in the atmosphere not only is a good indicator of water-splitting photosynthesis but may be vital for the development of any life more complex than single-celled bacteria. Aerobic respiration releases more energy than any other redox process except the reduction of fluorine or chlorine. However, these two gases are far too reactive ever to accumulate in an atmosphere, whereas oxygen is relatively stable. The cells of complex organisms like animals, plants and fungi consume ten times more power than prokaryotic cells and so require a very rich source of energy. An oxygen-laden atmosphere is possibly the only way more complex life could support its power demands.

So, if spectroscopic analysis shows an atmosphere far from chemical equilibrium, for example one containing high levels of both oxygen and methane, astrobiologists will have a good case for suspecting life. Carbon dioxide, water vapour and methane all give good signals to an infra-red spectroscope; excellent absorption of infra-red is the very reason such molecules act as effective green-house gases. Atmospheric oxygen cannot itself be easily identified by infra-red spectroscopy but UV in sunlight converts it into ozone which does absorb well.

None the less, the absence of a clear-cut disequilibrium is not necessarily evidence against the presence of life. Although a globally-distributed surface ecosystem leaves its signature on the atmosphere of the planet, that of less widely-established life may remain undetectable. Possible organisms on Mars may have retreated to isolated refuges deep underground. Although trace amounts of methane have been detected there, such a slight signal of life on extra-solar planets would be impossible to detect. Even on Earth, it would have taken about a billion years of photosynthesis before atmospheric oxygen could have been detected.

These biosignatures have already been spotted in the atmosphere of a planet. In 1990, the *Galileo* probe swung past the Earth for a gravitational boost to send it *en route* to Jupiter. As it raced away, it turned its instruments back and measured the spectrum of reflected sunlight. *Galileo* detected the tell-tale dips indicating the presence of oxygen and methane in extreme chemical disequilibrium. The probe also saw the absorption fingerprint of chlorophyll, the mole-cule that captures sunlight during photosynthesis. All these signs, taken together, allowed *Galileo's* operators to assert, with almost complete conviction, that the Earth does in fact harbour life.

More recent research has looked at the spectrum of Earthshine – the light coming off the Earth reflected back by the Moon. This diffuse light more closely resembles what we will be able to collect from extra-solar planets and makes a good test of analytical skills. In Earthshine, the astronomers were able to detect the spectral signature of water and ozone and also the 'red cliff' of vegetation. Plants absorb visible light only down to a certain frequency and reflect all infra-red light, to avoid overheating. If our eyes were sensitive to this portion of the spectrum we would see forests and grassland vivid infra-red rather than bright green. Viewed on the electromagnetic spectrum, vegetation causes a sudden jump in absorbance beyond the infra-red, hence the nickname the 'red cliff'. The frequency the red cliff occurs at is particular to the cellular structure of vegetation; not all terrestrial land plants produce a cliff at the same point. Although some astrobiologists believe the chlorophyll molecule could be employed by life throughout the galaxy – its building blocks are found in the interstellar dust clouds – there is no certainty that alien forests would be detectable by a similar spectral feature. Photosynthesis under a red dwarf star would need a leaf structure and chlorophyll-like molecule tuned to lower-energy light and able to harness three or four particles of light to release each electron. A more universal biosign for vegetation, therefore, would be the observation of any kind of spectral cliff, the magnitude of which waxes and wanes with the seasons on the planet.

Face of an alien world

Looking beyond space telescopes such as *TPF* and *Darwin*, astronomers have even more ambitious plans. The resolving power of such an interferometer is limited by the baseline distance across the array but once the technological challenges have been mastered the arrays can be arbitrarily large. Why settle for a virtual telescope 100 m in diameter, when you could spread your mirrors across hundreds or even thousands of kilometres? Such an enormous array would have a resolution good enough to make out features on the surface of a distant terrestrial planet. This speculative project, the *Planet Imager*, could perhaps reach completion within thirty years. The low-resolution images produced might look like a Pointillist abstraction of dots but the sight of a terrestrial extra-solar planet would be truly spectacular. We might see a spread of continents, wide oceans and bright polar ice caps. If the planet has life, we could see desert equatorial regions contrasting against the colour of plains and forests (they may even show the same red cliff as our own plant-powered ecologies). We might see seasonal fluctuations in the vegetation on a continent as the hemisphere swings between summer and winter.

Imagine that; gazing on the face of an alien earth a mere 425 years after Galileo first turned his telescope to the heavens.

8

Synthesis

Throughout this book, I have focused almost exclusively on prokaryotic life. Compared to our own eukaryotic cells, these primitive organisms are fantastically hardy and can use an extremely broad range of energy sources. Eukaryotes have existed for about half the history of life on our planet; complex animals and land plants for merely the last sixth. As extra-terrestrial life, simple prokaryotic-like cells are by far the most promising. Such life may one day be found swarming in a great variety of different cosmic habitats; on planets huddling close to their red dwarf star, on the small moons of gas giants or high in the clouds of green-house worlds. But complex animal life requires much more stringent and stable conditions and may only be possible around Sun-like stars, on very Earth-like planets with plate tectonics, oceans of water, continental land, a thick oxygen-rich atmosphere and large moon.

What would an alien look like?

Assuming conditions on a far-flung world were suitable for complex life – multi-cellular plants and animals – what might that life be like? In this final chapter I shall allow my imagination to run a little more freely and to speculate (from solid scientific foundations) how complex alien life on a terrestrial planet might have evolved. This is essentially the same as asking what would be produced if evolution were reset on Earth. What would happen if the tape of terrestrial life were rewound to the beginning and set to run again? Could you expect the results, four billion years later, largely to correspond to today's ecosystems or would the world be populated by completely unrecognisable organisms?

There may be very few ways to skin this proverbial cat. It is possible the system of RNA/DNA and proteins enveloped in a fatty membrane is the only workable way of producing life from organic substances. If this is so, aliens on a terrestrial planet could be biochemically very similar to life on ours. Once the first cells have arisen, the same evolutionary pressures will apply, forcing life to exploit all possible energy sources; to compete for space, nutrients or other resources or co-operate for mutual gain. Some researchers also consider the symbiotic development of a eukaryotic-like cell and the building of multi-cellular organisms to be all but inevitable once life gets started.

In many respects, evolution hasn't been inspirationally original over the billions of years that separate humans from bacteria but it has been extraordinarily cunning in rejigging existing proteins into new functions. Mother Nature is primarily a tinkerer rather than an inventor, re-using the bacterial tool-kit to jury-rig solutions to new problems. For example, the tunnel-like proteins that allow impulses to travel along our brain cells are directly related to the channels spanning bacterial membranes. All human thought from language and philosophical pondering to the design of the Space Shuttle, is based on an ancient molecular technology developed by bacteria to regulate their internal environment.

Furthermore, it seems that evolution is not free to explore all the possible designs for an organism but is constrained to follow a particular course, like water flowing along a steep gorge. There are only a few feasible ways of solving a particular survival problem and evolution hits upon the same designs again and again. For example, within the animal kingdom, eyes have independently evolved a great number of times, from primitive light-sensitive pits, to insect compound eyes and the camera eyes of vertebrates such as humans. The vertebrate eye is a marvel of biomechanical engineering, with an adaptive lens and dynamic aperture producing crystal-clear vision in wide-ranging conditions. But this camera-design eye has evolved independently at least five times in unrelated groups of animals. Natural selection, acting on completely different organisms, has converged on the same solution. The cephalopod eye (like that of octopus and squid) is actually far superior to the vertebrate eye, in that it has no blind spot where the optic nerve leaves the eyeball. If life were restarted from scratch, you would expect animals soon to develop sight, if not necessarily even a camera-eye, because it confers such an enormous survival advantage in sensing the environment.

One of the most important questions in evolutionary biology is what features of organisms are universal and might be expected to re-appear no matter how many times life were restarted. Structures such as the eye, that have independently evolved a number of times, are good candidates; we might expect aliens on a similar planet also to bear these adaptations. Other aspects of organisms are inconsequential, historical contingencies that could easily have turned out differently: why do humans have five digits on each hand? Evolutionarily, five might be no more advantageous than four or six. Some evolutionary biologists think fiveness might have been an essentially random outcome; it's just that fish with five bones supporting their fins beat their relatives on to dry land. Had a different fishy ancestor scrambled up first, we might be counting in base 8 (8, 64, 512) rather than base 10 (10, 100, 1000). Other random events can also have enormous effects on ecosystems, such as the meteorite impact that knocked out seventy-five per cent of species 65 million years ago.

Convergence happens because diverse organisms are subject to the same physical laws. Many marine species, including fish like sharks, mammals like whales and dolphins, extinct dinosaurs such as the ichthyosaur and, to a certain degree, birds like the penguin, all have a tapered, smooth-skinned, body, because this is an efficient, streamlined shape. Any alien species selected for fast swimming would also have to look like this, regardless of whether it swam through liquid water or liquid ethane.

The evolution of flight may also be inevitable on terrestrial planets. On Earth, it has evolved in the extinct pterosaurs, insects, birds and bats. Certain amphibians, some fish and to a limited extent the seeds of some plants, glide. Although the energetics of staying airborne impose severe limits on the size of flying animals, it still has enormous benefits in evading predators or foraging for food. Some interesting calculations have been performed on how flight might be possible on other worlds. Planets smaller than Earth would have a lower gravity and so tend to have thinner atmospheres. This means that less lift can be generated by wings and flight would be difficult. Wings with very large surface areas, which flapped in a gentle rhythm, would be needed, probably much like those of the dragonflies with 75 cm wing-spans that filled Earth's skies 300 million years ago. Paradoxically, flight would be easier on a planet with a stronger gravitational pull. A much thicker atmosphere near the surface would mean even modestly-sized wings could generate great lift forces.

Other features of Earthly animals are thought to be such effective adaptations that their structure ought to recur regularly on other worlds. One primitive body plan is a long tube, made up of repeating identical, and largely independent, sections. This segmented construction is relatively easy for a genetic system to encode, growth is as simple as adding extra segments to the end and movement can be achieved by a simple wave motion passing down the length of the animal. This body plan is obvious in worms and still evident as an evolutionary relic in the three-part body of insects and even in our own repeated vertebrae and ribs.

An efficient way of processing food and extracting nutrients is by flow along a pipe-like gut with an 'entrance' and 'exit'. (The human body plan, as that of all other vertebrates, is effectively no more than an elaborated gut tube, supported by a backbone and accessorised with limbs.) Transporting nutrients to where they're needed in a large animal, and removing waste, is best achieved by an internal circulation system. Vertebrates use the iron-containing molecule, haemoglobin, to carry oxygen to their metabolising cells and so have red blood but other metal ions can also be used; horseshoe crabs are blue-blooded with copper while sea cucumbers use a yellow-green pigment.

Gathering information about your surroundings helps you find food and avoid predators and so the evolution of appropriate senses is to be expected. Hearing, or sensing vibrations, helps in most environments; vision is useful if you live within the sunlit zone; chemoreception (taste and smell) reports on important chemicals and some animals are even sensitive to magnetic and electric fields. A centralised nervous system allows faster processing of all this information and it is sensible to keep your main sensory organs near it, to minimise reaction times. Thus the development of a head at the front of the body might be universal amongst higher animals, although the particular positioning of eyes, ears, nose and mouth would not be.

Convergence is also prevalent in the plant world. Plants optimise five basic requirements: water conservation, exchange of gases with the atmosphere, sunlight catchment, spore dispersal and mechanical stability. Depending on the demands of the local environment, plants find different compromises. Cactuses, in arid regions, are stubby, with few offshoots, to conserve water, whereas plants maximising light interception have wide, flat canopies. Similar constraints would operate on an alien world and so the shape of trees might be remarkably familiar. On a planet with stronger gravity or high winds, mechanical stability would be the prime consideration and trees would be low-lying with little sideways spread of branches. On a planet with extreme winds, such as one tidally locked to a red dwarf star, trees might resemble seaweed, which survives the battering of breaking waves with a submissively flexible stem and streaming fronds.

In many ways, complex alien life may closely resemble Earth's. We would instantly recognise trees and even the animals should have identifiable features such as segments, fins or limbs, eyes, guts and hearts. The arrangement of sense organs, skin patterning and the number of legs are more likely to be contingent and so not predictable.

The question of whether human-level intelligence is a contingent or an inevitable product of evolution is fiercely debated and not a topic that I shall cover. But let us assume that an extra-terrestrial species evolves to be at least as intelligent as humankind and (in homage to a 1950s B-movie) lands his or her (but let's not get into the headache of what sex an alien might be, assuming they indeed have two sexes or even reproduce sexually) brightly-polished flying saucer on the lawn of the White House. What might we expect the aliens that step, crawl, slither or ooze out of the hissing air lock to look like? Are they really likely to be green, bug-eyed monsters?

We can make certain reasonably confident assertions. ET would almost certainly need to be a land animal. The fins needed for marine life, like those of fish or dolphins, are not particularly dextrous and are unsuitable for manipulating tools to build artefacts or modify the environment. Although nimble, octopus-like creatures are an obvious counter to this thought, it is difficult to imagine how fire might be exploited underwater and smelting metals from their ores was crucial for the development of human technology.

Might our visitor be green? This colouration, in terrestrial organisms, is usually for one of two reasons: in plants it derives from the photosynthetic pigment chlorophyll, in animals from the need to be camouflaged to remain hidden in this vegetation. Some researchers think chlorophyll may be a universal molecule, used by organisms throughout the galaxy to soak up the light of their star. However, the wavelength of light emitted by a star depends on how hot it is, so a star cooler than our Sun will shine more in the infra-red than the visible portion of the spectrum. Photosynthetic pigments would therefore need to be differently tuned to absorb the light from a red dwarf and would not look green to our eyes. Thus, an alien with a cooler home star could be well-camouflaged to the colour of their vegetation but would not appear green to us. The alien's eyes would be most sensitive to this lower wavelength light, so they would perceive the world in a very different way, chosing infra-red 'colours' to paint their space ship. If the beings stepping out of the flying saucer look bright green you can therefore either attribute it to an unlikely coincidence or conclude that their sun is much like our own.

Or perhaps the alien's skin is green not for camouflage against vegetation but because it contains chlorophyll to supply nutrients. Earth plants are self-sufficient through photosynthesis but animals must get their energy indirectly, either by eating vegetation or eating another animal that has. Given a fresh start, what's stopping evolution endowing lizards with green solar panel-skin to generate a little energy during the day? The simple answer is that photosynthesis, even assuming the animal never needs to hide from predators nor get into the shade for any other reason, cannot provide nearly enough energy to meet an animal's demands. Only about eight per cent of the light energy falling on a plant is converted to sugars, which for the skin area of a roughly human-sized animal doesn't provide even a tenth of the energy required by the muscles and brain. Complex animal life cannot support itself with photosynthesis but must harvest the concentrated ready-made nutrients contained within plants or other animals and will almost certainly need to respire with oxygen to release enough energy for the high demands of their cells.

There are also good evolutionary reasons not to expect the visitor to be bug-eyed. Sight is an extremely important sense for perceiving one's surroundings and is thought likely to be a universal characteristic of complex land animals. But many of the alternative eye designs that have evolved give very poor visual acuity. They may be fine for seeing which direction the Sun is in or spotting a lunging predator in time but do not provide sharp enough resolution to spur the development of high intelligence. The compound eye of insects is thought to provide reasonably clear vision with a wide field of coverage but it has been calculated that a compound eye with an acuity equal to that of humans would need to be over a metre across. There is every chance, therefore, that ET will gaze back at us with camera-type eyes like our own.

What about the technological alien being a 'monster'. For the same energetic constraints described above, any highly intelligent animal probably needs to be, at least partly, carnivorous. Big brains (or whatever centralised mass of nerve cells ET has) have expensive power demands, which may necessitate the high returns gained from consuming other animals. Active hunters also experience evolutionary pressure towards greater processing abilities to outwit their prey, whereas herbivores spend much of their time idly harvesting plants. Our alien travellers couldn't be a species of fiercely territorial, solitary hunters. They would need to be a social species, able to live relatively peacefully in dense populations, to allow cultural learning, the exchange of information and ideas, to accumulate knowledge over generations and to allow technology to advance. A space ship cannot be built by a species of isolated individuals but only by the co-operative efforts of a society. Furthermore, a very belligerent species is likely to destroy itself before spreading through the galaxy. On our own path of progress, the technological capability for space-faring or interstellar radio communication came at almost exactly the same time we discovered the extraordinary properties of the ninety-second element, uranium. After realising its potential for undergoing runaway nuclear fission we immediately put it to good use in building weapons capable of obliterating our entire species. Thus, a species of aggressive monsters is likely to kill itself off as soon as it becomes technological, long before being able to voyage between the stars and land on the White House lawn.

The creatures that emerge from the air-lock are unlikely to be the two-armed upright bipeds, with recognisable faces, favoured by sci-fi shows for the sake of their costume department. But they will probably share some universal features with us, such as a gut, a system of internal vessels and a head with eyes. They will almost certainly have evolved from a segmented ancestor and will enjoy a nice deep lung-full of oxygen as much as the next space-faring alien.

Conclusion

By 1991, the Voyager 1 space probe had travelled over four billion miles on its grand tour through the solar system and the NASA scientists set it one last photographic assignment. As the aging probe neared the edge of the solar system, it turned back towards the feeble light of its distant home star and photographed each of the planets in turn, creating a family portrait of the Sun and her brood. As shown in Figure 19, seen from such a great distance the Earth barely fills a single pixel of the image; it is no more than a 'pale blue dot' adrift in the vast darkness of space (the bright streak crossing the frame is sunlight catching in the lens). This lone point of light encompasses our entire world, over six billion humans, countless trillions of other animals, plants, fungi and a mind-bogglingly large number of prokaryotes, all inhabiting the same lump of rock, producing, devouring, parasitising, competing but ultimately mutually interdependent, inextricably linked together into enormously complex ecosystems. This speck in the cosmos is our home and the only known oasis of life in the entire universe.

Over the pages of this book, we've taken a close look at this pale blue dot and the emergence, spread and development of life on it over the billions of years of its history. The central drive of astrobiology is not only to understand the origins and processes of life but also how they are dependent on the physical and chemical environment of our world and its cosmic neighbourhood. Life is demonstrably possible on Earth but how feasible is it on other Earth-like planets such as Mars, on those orbiting other stars throughout the galaxy or even in exotic situations very unlike our own? Astrobiology is a very young science and is only just

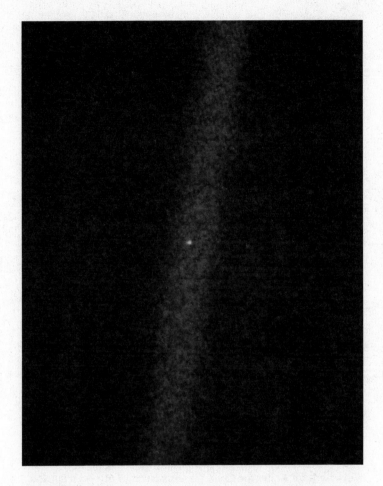

Figure 19 A pale blue dot adrift in the dark void – the Earth from over 4 billion miles away. Voyager 1 turns its cameras back on our world one last time as it exits the solar system.

beginning to answer the multitude of questions presented in this book. The huge unexplored territory and unknowns of astrobiology make it at times a bewildering and frustrating pursuit. But over the next twenty years it will certainly mature into one of the most exciting areas of scientific research.

Other than on Earth, there is certainly no complex animal or plant life within our solar system. Although I've discussed several other possible habitats suitable for single-celled organisms, each has its associated problems. Venus has no habitable surface and only a narrow cloud layer offers tolerable conditions; Europa has a water ocean but possibly lacks a dependable energy source; Titan has available chemical energy but may be insurmountably cold. Many astrobiologists consider the most promising candidate to be Mars but life may thrive there only deep beneath the freeze-dried surface. Where will astrobiologists first find convincing evidence of extra-terrestrial life? My belief is that surveys of extra-solar planets will find the tell-tale signature of chemical disequilibrium in the atmosphere of an alien earth long before we have been able to explore extensively, with robotic probes, the possible subsurface habitats of Mars or Europa in our solar system.

Surely the greatest enduring mystery of biology is the origin of life on Earth; understanding how you get from ubiquitous simple organic substances to replicating polymers and metabolisms. We really don't have much of a clue about the how, where, when and how long. Defining the range of conditions that would permit life on other worlds is even harder. Lab experiments are useful but they can only provide a list of possible mechanisms rather than demonstrating the processes that actually occurred four billion years ago. Even if we succeed in recreating life from scratch in a test tube, we will still never know the exact circumstances of our own genesis.

The fundamental question is whether life will be found 'here, there and everywhere' or are our nearest neighbours residing in a 'galaxy far far away'? Some researchers believe that the emergence of life is effectively inevitable, given appropriate conditions; indeed there is evidence that life arose almost instantaneously on our own world. Supporters point to examples where order

Figure 20 The Hubble Deep Field image. Although this photograph of the sky covers only 1% of the area of the full Moon it still contains hundreds of individual galaxies.

emerges spontaneously from chaotic systems and patterns self-assemble and drive towards increasing complexity. The emergence of life may not be due to a lucky concatenation of unlikely events but simply be a consequence of the natural development of organic chemistry, like the networks of autocatalytic reactions I described earler. Others believe that the power of such emergent systems is grossly over-stated and the development of life is extremely improbable. Having only one case to work from, it is impossible to estimate the likelihood of life arising but Earth could be the only inhabited world in our neighbourhood. However, considering the mind-boggling size of the visible universe, it is tempting to argue that life *must* exist elsewhere, just perhaps far beyond our reach. The image in Figure 20 illustrates well the sheer amount of real estate available. Even the most distant galaxies in the visible universe contain the same simple molecules found in the interstellar dust clouds of our own galaxy and so are probably also rife with life-building organic substances.

The chances are the whole universe is teeming with life, we just need to start reaching out to find it.

Glossary

AU

Astronomical Unit. The standard measure of distances within the solar system. It is defined as the mean orbital distance of the Earth from the Sun and is nearly 150 million km.

autotroph

An organism able to synthesise all the complex organic molecules it requires directly from inorganic sources such as carbon dioxide. This includes all photosynthetic plants and algae and many prokaryotes. See heterotroph.

bombardment

The period soon after the formation of the solar system when the inner planets were subjected to many impacts. It is not clear whether this was a tailing-off after the birth of the planets or a renewed blitz after a period of calm.

chemosynthesis

The method of deriving metabolic energy and producing organics from inorganic redox reactions. See photosynthesis.

eukaryote

A cell internally compartmentalised with membrane-bound organelles, specifically a nucleus. Includes all animals, plants, algae, fungi and some unicellular organisms. See prokaryote.

extremophile

An organism able to survive in an environment considered to be physically or chemically extreme and so hostile. Most known extremophiles are prokaryotic.

heterotroph

An organism that cannot produce its own complex organic molecules and so must consume other organisms or their remains. This includes all animals, as well as many other eukaryotes and prokaryotes. See autotroph.

HZ

Habitable Zone. A region of space where conditions are believed to be favourable for life. This can be either a stellar habitable zone or a galactic habitable zone.

ion

Any atom or molecule containing an unbalanced number of electrons, thus carrying a negative or positive electrical charge.

light year (ly)

A standard measure of astronomical distances. It is the distance travelled by light over the course of one year, equivalent to almost 9.5 trillion km.

LUCA

Last Universal Common Ancestor (also called Most Recent Common Ancestor, MRCA). The cell type or collective of cells from which all life on Earth is descended.

M-class dwarf

A very common type of small and cool 'red dwarf' star.

metabolism
The total network of biochemical reactions performed by a cell.

metallicity
The proportion of an object made up of chemical elements heavier than hydrogen and helium (such elements are termed metals by astronomers).

Milky Way
Our galaxy.

organic
A large class of molecules containing carbon that are vital for life as we know it but are not necessarily biologically-produced.

panspermia
The theory that life can be transferred between planets and moons within meteorites.

photosynthesis
The method of deriving metabolic energy and producing organics by using light energy. See chemosynthesis.

prokaryote
A primitive cell type without membrane-bound organelles and specifically lacking a nucleus. Includes all archaea and bacteria. See eukaryote.

redox
A chemical process comprised of both reduction (whereby a chemical gains an electron) and oxidation (whereby a chemical loses an electron) reactions.

snow line

The boundary within the accretion disc of a young star beyond which it is cool enough for volatiles to condense. Terrestrial planets form within this cut-off and so are originally very poor in volatiles essential for life, including water. Beyond the snow line large ice-containing planetary embryos form gas giant planets.

terrestrial

Referring either specifically to the Earth or more generally a small rocky Earth-like planet, including the other inner planets of the solar system, Mercury, Venus and Mars.

UV

Ultraviolet. A form of energetic light lying just beyond the visible portion of the electromagnetic spectrum.

volatiles

Small light molecules that make up atmospheres and seas, such as carbon dioxide, ammonia and water and are also vital for building the organic molecules needed for life.

Further reading

Hopefully you've found this Beginner's Guide to Astrobiology interesting and useful and have had your appetite whetted for more. I've listed below some other good books on this fascinating subject, arranged vaguely into an order of popular science to textbook. Please note, though, that none of the suggested reading lists presented here are intended to exhaustively cover the field or provide a complete bibliography of references used in this book, they are simply here to help you continue your exploration.

Life Everywhere: The Maverick Science of Astrobiology. David Darling

If the Universe Is Teeming with Aliens – Where Is Everybody?: Fifty Solutions to Fermi's Paradox and the Problem of Extra-terrestrial Life. Stephen Webb

The Extra-terrestrial Encyclopaedia. David Darling

The Chemistry of Life (Penguin Press Science). Steven Rose

Life's Solution: Inevitable Humans in a Lonely Universe. Simon Conway Morris

The Biological Universe: The Twentieth Century Extra-terrestrial Life Debate and the Limits of Science. Steven J. Dick

Search for Life on Other Planets. Bruce Jakosky

Life in the Solar System and Beyond. Barrie William Jones

Life in the Universe. Jeffrey Bennett, Seth Shostak, Bruce Jakosky

Astrobiology: The Quest for the Conditions of Life. G. Horneck, C. Baumstark-Khan (eds.)

Life in the Universe: Expectations and Constraints (Advances in Astrobiology & Biogeophysics). Louis N Irwin, Dirk Schulze-Makuch

Intelligent Life in the Universe: Principles and Requirements Behind Its Emergence (Advances in Astrobiology & Biogeophysics). Peter Ulmschneider

The following websites are well worth a browse as they have a good database of articles and also offer a regular news email.

http://www.astrobio.net/

http://www.astrobiology.com/

http://www.seti.org
http://www.space.com/

It is also worth subscribing to the email service offered by science magazines such as New Scientist and Scientific American as these often run articles of astrobiological interest. In terms of the primary academic literature, these journals commonly contain relevant papers:

Annual Review of Earth and Planetary Sciences
Astrobiology
Icarus
Journal of Geophysical Research
Nature
Origins of Life and Evolution of Biospheres
Planetary and Space Science
Science
Space Science Reviews

I've also compiled a short selection of some particularly interesting papers:

Ball, P. (2005). "Water and life Seeking the solution." *Nature* **436**(7054): 1084.

Banin, A. and Mancinelli, R. L. (1995). "Life on Mars? I. The chemical environment." *Advances in Space Research* **15**(3): 163.

Bartel, D. P. and Unrau, P. J. (1999). "Constructing an RNA world." *Trends in Biochemical Sciences* **24**(12): M9.

Beatty, J. T., Overmann, J., et al. (2005). "An obligately photosynthetic bacterial anaerobe from a deep-sea hydrothermal vent." *PNAS* **102**(26): 9306–9310.

Beier, M., Reck, F., et al. (1999). "Chemical Etiology of Nucleic Acid Structure: Comparing Pentopyranosyl-(2'4') Oligonucleotides with RNA." *Science* **283**(5402): 699–703.

Boston, P. J., Ivanov, M. V., et al. (1992). "On the possibility of chemosynthetic ecosystems in subsurface habitats on Mars." *Icarus* **95**(2): 300.

Canup, R. M. and Asphaug, E. (2001). "Origin of the Moon in a giant impact near the end of the Earth's formation." *Nature* **412**(6848): 708.

Catling, D. C., Glein, C. R., et al. (2005). "Why O2 is required by complex life on habitable planets and the concept of planetary "oxygenation time"." *Astrobiology* **5**(3): 331–332.

Cavalier-Smith, T., Brasier, M., et al. (2006). "Introduction: how and when did microbes change the world?" *Phil. Trans. R. Soc. B* **361**: 845–850.

Chapelle, F. H., O'Neill, K., et al. (2002). "A hydrogen-based subsurface microbial community dominated by methanogens." *Nature* **415**(6869): 312.

Chyba, C. (1997). "Life on other Moons." *Nature* **385**: 201.

Chyba, C. F. (1990). "Impact delivery and erosion of planetary oceans in the early inner Solar System." *Nature* **343**: 129–133.

Chyba, C. F. (2000). "Energy for microbial life on Europa." *Nature* **403**(6768): 381.

Chyba, C. F. and McDonald, G. D. (1995). "The Origin of Life in the Solar System: Current Issues." *Annual Review of Earth and Planetary Sciences* **23**: 215–249.

Chyba, C. F. and Sagan, C. (1992). "Endogenous production, exogenous delivery and impact-shock synthesis of organic molecules: an inventory for the origins of life." *Nature* **355**: 125–132.

Clark, B. C. (1998). "Surviving the limits to life at the surface of Mars." *Journal Of Geophysical Research-Planets* **103**(E12): 28545–28555.

Cockell, C. and Lee, P. (2002). "The biology of impact craters – a review." *Biological Reviews* **77**: 279–310.

Cockell, C., Lee, P., et al. (2002). "Impact-induced microbial endolithic habitats." *Meteoritics & Planetary Science* **37**: 1287–1298.

Cockell, C. S. (1999). "Life on Venus." *Planetary and Space Science* **47**(12): 1487–1501.

Cohen, J. and Stewart, I. (2001). "Where are the dolphins?" *Nature* **409**(6823): 1119–1122.

Cooper, G., Kimmich, N., et al. (2001). "Carbonaceous meteorites as a source of sugar-related organic compounds for the early Earth." *Nature* **414**(6866): 879.

Doolittle, W. F. (1999). "Phylogenetic Classification and the Universal Tree." *Science* **284**(5423): 2124–2128.

Duboule, D. and Wilkins, A. S. (1998). "The evolution of 'bricolage'." *Trends in Genetics* **14**(2): 54.

Ellis, J. and Schramm, D. N. (1995). "Could a Nearby Supernova Explosion have Caused a Mass Extinction?" *PNAS* **92**(1): 235–238.

Eschenmoser, A. (1999). "Chemical Etiology of Nucleic Acid Structure." *Science* **284**(5423): 2118–2124.

Ford, E. B., Seager, S., et al. (2001). "Characterization of extra-solar terrestrial planets from diurnal photometric variability." *Nature* **412**(6850): 885–887.

Formisano, V., Atreya, S., et al. (2004). "Detection of Methane in the Atmosphere of Mars." *Science* **306**(5702): 1758–1761.

Fortes, A. D. (2000). "Exobiological Implications of a Possible Ammonia–Water Ocean inside Titan." *Icarus* **146**(2): 444–452.

Franck, S., Block, A., et al. (2000). "Habitable zone for Earth-like planets in the solar system." *Planetary and Space Science* **48**(11): 1099.

Gaidos, E. J., Nealson, K. H., et al. (1999). "BIOGEOCHEMISTRY:Life in Ice-Covered Oceans." *Science* **284**(5420): 1631–1633.

Gies, D. R. and Helsel, J. W. (2005). "Ice Age Epochs and the Sun's Path through the Galaxy." *The Astrophysical Journal* **626**: 844–848.

Gold, T. (1992). "The deep, hot biosphere." *Proceedings of the National Academy of Science USA* **89**(13): 6045–6049.

Gonzalez, G., Brownlee, D., et al. (2001). "The Galactic Habitable Zone: Galactic Chemical Evolution." *Icarus* **152**(1): 185–200.

Greaves, J. S. (2005). "Disks Around Stars and the Growth of Planetary Systems." *Science* **307**(5706): 68–71.

Hiscox, J. A. (2001). "An Overview Of The Origin Of Life: The Case For Biological Prospecting On Mars." *Earth, Moon and Planets* **87**(3): 191.

Hoffman, P. F., Kaufman, A. J., et al. (1998). "A Neoproterozoic Snowball Earth." *Science* **281**(5381): 1342–1346.

Horneck, G. (2000). "The microbial world and the case for Mars." *Planetary and Space Science* **48**(11): 1053.

Irwin, L. N. and Schulze-Makuch, D. (2001). "Assessing the plausibility of life on other worlds." *Astrobiology* **1**(2): 143.

Kaiser, D. (2001). "Building a multi-cellular organism." *Annual Review of Genetics* **35**(1): 103–123.

Kargel, J. S. (2004). "PLANETARY SCIENCE: Enhanced: Proof for Water, Hints of Life?" *Science* **306**(5702): 1689–1691.

Kasting, J. F., Whitmire, D. P., et al. (1993). "Habitable Zones around Main Sequence Stars." *Icarus* **101**: 108–128.

Kerr, R. A. (2004). "BREAKTHROUGH OF THE YEAR: On Mars, a Second Chance for Life." *Science* **306**(5704): 2010–2012.

Knoll, A. H. and Carroll, S. B. (1999). "Early Animal Evolution: Emerging Views from Comparative Biology and Geology." *Science* **284**(5423): 2129–2137.

Konacki, M. (2005). "An extra-solar giant planet in a close triple-star system." *Nature* **436**(7048): 230.

Lazcano, A. and Miller, S. L. (1996). "The Origin and Early Evolution of Life: Prebiotic Chemistry, the Pre-RNA World and Time." *Cell* **85**(6): 793.

Leitch, E. M. and Vasisht, G. (1998). "Mass extinctions and the sun's encounters with spiral arms." *New Astronomy* **3**(1): 51.

Line, M. A. (2002). "The enigma of the origin of life and its timing." *Microbiology* **148**(1): 21–27.

Lineweaver, C. H., Fenner, Y., et al. (2004). "The Galactic Habitable Zone and the Age Distribution of Complex Life in the Milky Way." *Science* **303**(5654): 59–62.

Lipps, J. H., Delory, G., et al. "Astrobiology of Jupiter's Icy Moons." *preprint*.

Lissauer, J. J. (1999). "How common are habitable planets?" *Nature* **402**(6761): C11.

Lissauer, J. J. (2002). "Extra-solar planets." *Nature* **419**(6905): 355–358.

Lovelock, J. E. (1975). "Thermodynamics and the Recognition of Alien Biospheres." *Proceedings of the Royal Society of London. Series B, Biological Sciences* **189**(1095, A Discussion on the Recognition of Alien Life): 167–180.

Mancinelli, R. L. and Banin, A. (1995). "Life on Mars? II. Physical restrictions." *Advances in Space Research* **15**(3): 171.

Martin, W. and Embley, T. M. (2004). "Early evolution comes full circle." *Nature* **431**: 134–135.

Mastrapa, R. M. E., Glanzberg, H., et al. (2001). "Survival of bacteria exposed to extreme acceleration: implications for panspermia." *Earth and Planetary Science Letters* **189**(1–2): 1.

McKay, C. P. and Smith, H. D. "Possibilities for methanogenic life in liquid methane on the surface of Titan." *Icarus* **178**(1): 274–276.

McKay, D. S., Gibson, E. K. Jr., et al. (1996). "Search for past life on Mars: possible relic biogenic activity in Martian meteorite ALH84001." *Science* **273**(5277): 924.

Mileikowsky, C., Cucinotta, F. A., et al. (2000). "Natural Transfer of Viable Microbes in Space. 1. From Mars to Earth and Earth to Mars." *Icarus* **145**: 391–427.

Mileikowsky, C., Cucinotta, F. A., et al. (2000). "Risks threatening viable transfer of microbes between bodies in our solar system." *Planetary and Space Science* **48**(11): 1107.

Moorbath, S. (2005). "Palaeobiology: Dating earliest life." *Nature* **434**(7030): 155.

Nealson, K. H. (1997). "The limits of life on Earth and searching for life on Mars." *Journal Of Geophysical Research* **102**(E10): 23,675–23,686.

Nealson, K. H., Inagaki, F., et al. (2005). "Hydrogen-driven subsurface lithoautotrophic microbial ecosystems (SLiMEs): do they exist and why should we care?" *Trends in Microbiology* **13**(9): 405.

Newsom, H. E., Brittelle, G. E., et al. (1996). "Impact crater lakes on Mars." *Journal Of Geophysical Research-Planets* **101**(E6): 14951–14955.

Niklas, K. J. (2004). "Computer Models of Early Land Plant Evolution." *Annual Review of Earth and Planetary Sciences* **32**: 47–66.

Nisbet, E. G. and Sleep, N. H. (2001). "The habitat and nature of early life." *Nature* **409**(6823): 1083–1091.

Nyquist, L. E., Bogard, D. D., et al. (2001). "Ages and Geologic Histories of Martian Meteorites." *Space Science Reviews* **96**(1–4): 105.

Orgel, L. (2000). "ORIGIN OF LIFE: Enhanced: A Simpler Nucleic Acid." *Science* **290**(5495): 1306–1307.

Orgel, L. E. (1998). "The Origin of Life – How Long did it Take?" *Origins of Life and Evolution of Biospheres* (Formerly *Origins of Life and Evolution of the Biosphere*) **28**(1): 91.

Orgel, L. E. (1998). "The origin of life–a review of facts and speculations." *Trends in Biochemical Sciences* **23**(12): 491.

Pedersen, K. (2000). "Exploration of deep intraterrestrial microbial life: current perspectives." *FEMS Microbiology Letters* **185**(1): 9.

Perryman, M. A. C. (2000). "Extra-solar planets." *Reports on Progress in Physics* (8): 1209.

Porco, C. C., Baker, E., et al. (2005). "Imaging of Titan from the Cassini spacecraft." *Nature* **434**(7030): 159.

Raulin, F. and McKay, C. P. (2002). "Special issue on Exobiology: the search for extra-terrestrial life and prebiotic chemistry." *Planetary and Space Science* **50**(7–8): 655.

Raulin, F. and Owen,T. (2002). "Organic Chemistry and Exobiology on Titan." *Space Science Reviews* **104**(1): 377–394.

Ravilious, K. (2005). Top 5 cosmic threats to life on Earth. *New Scientist*: 32–35.

Richardson, D. J. (2000). "Bacterial respiration: a flexible process for a changing environment." *Microbiology* **146**(3): 551–571.

Rothschild, L. J. and Mancinelli, R. L. (2001). "Life in Extreme Environments." *Nature* **409**: 1092–1101.

Rummel, J. D. (2001). "Special Feature: Planetary exploration in the time of astrobiology: Protecting against biological contamination." *PNAS* **98**(5): 2128–2131.

Sagan, C. and Salpeter, E. E. (1976). "Particles, environments and possible ecologies in the Jovian atmosphere." *The Astrophysical Journal Supplement Series* **32**: 737–755.

Schoning, K. U., Scholz, P., et al. (2000). "Chemical Etiology of Nucleic Acid Structure: The alpha -Threofuranosyl-(3' $\rightarrow$ 2') Oligonucleotide System." *Science* **290**(5495): 1347–1351.

Schrag, D. P., Hoffman, P. F., et al. (2001). "Life, geology and snowball Earth." *Nature* **409**: 306.

Schulze-Makuch, D. and Grinspoon, D. H. (2005). "Biologically enhanced energy and carbon cycling on Titan?" *Geochimica et Cosmochimica Acta Supplement* **69**: 528.

Schulze-Makuch, D. and Irwin, L. (2006). "The prospect of alien life in exotic forms on other worlds." *Naturwissenschaften* **93**(4): 155.

Schulze-Makuch, D. and Irwin, L. N. (2002). "Reassessing the Possibility of Life on Venus: Proposal for an Astrobiology Mission." *Astrobiology* **2**(2): 197–202.

Schulze-Makuch, D., Irwin, L. N., et al. (2002). "Search parameters for the remote detection of extra-terrestrial life." *Planetary and Space Science* **50**(7–8): 675.

Siegert, M. J. (2005). "LAKES BENEATH THE ICE SHEET: The Occurrence, Analysis and Future Exploration of Lake Vostok and Other Antarctic Subglacial Lakes." *Annual Review of Earth and Planetary Sciences* **33**(1): 215–245.

Simoneit, B. R. T., Summons, R. E., et al. (1998). "Biomarkers as Tracers for Life on Early Earth and Mars." *Origins of Life and Evolution of Biospheres* (Formerly *Origins of Life and Evolution of the Biosphere*) **28**(4–6): 475.

Stevens, T. O. and McKinley, J. P. (1995). "Lithoautotrophic Microbial Ecosystems in Deep Basalt Aquifers." *Science* **270**(5235): 450–455.

Thomas, D. N. and Dieckmann, G. S. (2002). "Antarctic Sea Ice–a Habitat for Extremophiles." *Science* **295**(5555): 641–644.

Waltham, D. (2004). "Anthropic Selection for the Moon's Mass." *Astrobiology* **4**(4): 460–468.

Wells, L. E., Armstrong, J. C., et al. (2003). "Reseeding of early earth by impacts of returning ejecta during the late heavy bombardment." *Icarus* **162**: 38–46.

Whitfield, J. (2004). "Exobiology: It's life. isn't it?" *Nature* **430**(6997): 288–290.

Williams, D. M., Kasting, J. F., et al. (1997). "Habitable Moons around extra-solar giant planets." *Nature* **385**: 234–236.

Wolstencroft, R. D. and Raven, J. A. (2002). "Photosynthesis: Likelihood of Occurrence and Possibility of Detection on Earth-like Planets." *Icarus* **157**(2): 535–548.

Woolf, N. J., Smith, P. S., et al. (2002). "The Spectrum of Earthshine: A Pale Blue Dot Observed from the Ground." *The Astrophysical Journal* **574**: 430–433.

Wynn-Williams, D. D. and Edwards , H. G. M. (2000). "Antarctic ecosystems as models for extra-terrestrial surface habitats." *Planetary and Space Science* **48**(11): 1065.

Index